Lecture Notes in Chemistry

Edited by G. Berthier M.J.S. Dewar H. Fischer
K. Fukui G.G. Hall H. Hartmann J. Hinze H.H. Jaffé J. Jortner
W. Kutzelnigg K. Ruedenberg E. Scrocco

36

I. Ugi
J. Dugundij
R. Kopp
D. Marquarding

Perspectives in Theoretical Stereochemistry

Springer-Verlag
Berlin Heidelberg New York Tokyo 1984

Authors

J. Dugundij
Department of Mathematics, University of Southern California
Los Angeles, CA 90089-1113, USA

R. Kopp
D. Marquarding †
I. Ugi
Organisch-Chemisches Institut der Technischen Universität München
D-8046 Garching

ISBN-13:978-3-540-13391-9   e-ISBN-13:978-3-642-93266-3
DOI: 10.1007/978-3-642-93266-3

Library of Congress Cataloging in Publication Data. Main entry under title: Perspectives in
theoretical stereochemistry. (Lecture notes in chemistry; 36) Includes bibliographies and index.
1. Stereochemistry. I. Ugi, Ivar, 1930-. II. Series.
QD481.P37 1984 541.2'23 84-14190
ISBN-13:978-3-540-10273-1 (U.S.)

2152/3140-543210

We dedicate this monograph to the memory of
Professor Dieter Marquarding,
who died on 9 July 1982 at the age of 47.
The theory presented here owes much to his effort:
his extensive knowledge and deep insight into the nature
of stereochemical processes was directly instrumental
in isolating and formulating many of our concepts.

The purpose of the mathematical physicists is not to calculate
phenomena quantitatively but to understand them qualitatively.
Their aim is to clarify with mathematical precision the meaning
of the concepts upon which physical theories are built.

Freeman J. Dyson
"Unfashionable Pursuits"
A. v. H. Stiftung
Mitteilungen *41*, 12 (1983)

Where is mathematical chemistry?

P R E F A C E

This treatment of stereochemistry was developed in numerous joint
discussions at the Technische Universität München over the period
1976-1982. It is applicable to all molecules, flexible or rigid,
and can be regarded as a complement to the known algebraic treatment of constitutional chemistry in terms of BE- and R-matrices.

We extend our gratitude to Dr. John Showell, who helped us formulate
and clarify some concepts; to Prof. Daniel S. Kemp, who critically
analyzed the manuscript and proposed numerous changes and additions
that improved the readability of this book; and to Profs. R. Bau,
M. Gielen, K. Mislow, F. Ramirez, K. Schäfer, Drs. J. Brandt, J.
Gasteiger, W. Schubert and Mr. T. Damhus for their helpful comments.
We wish to thank the Mss. Eva Nuytten, Sigrid Rössel, Herta Schön-
mann, Inge Schwarz, Marina Thoma, Maria Ulkan, and Mr. Michael
Capone for their patience and cheerful cooperation in the preparation, illustrations, revision, and proofreadings of the text, so
well as Dr. J. Bauer, Mr. E. Fontain and Mr. K. Stadler for the
development of computer software that was used in the production
of this manuscript.

The development of this book has gone through many stages and
versions over the years. We are very much indebted to Doz. Dr. Josef
Brandt and Ms. Sigrid Minker who went along all the way with us.
This monograph would never have reached the present form without
their creative contribution, diligence and patience in organizing,
computer-editing, improving and preparing in final form the manuscript, despite a variety of adverse conditions, including the
repeated breakdown of aged computer hardware.

We gratefully acknowledge the generous financial support given
to the project by the A. v. Humboldt Foundation, the Stiftung
Volkswagenwerk e.V., and the Fonds der Chemischen Industrie.

June, 1982   The Authors

INTRODUCTION

Stereochemistry is the part of chemistry that relates observable properties of chemical compounds to the structure of their molecules, i. e. the
relative spatial arrangement of their constituent atoms. In classical
stereochemistry, the spatial arrangements relevant for interpreting and
predicting a given chemical property are customarily described by geometric
features/symmetries in some suitably chosen rigid model of the molecule
[1].

The solution of stereochemical problems involving *single* molecular
species is the domain of the geometry based approaches, such as the methods
of classical stereochemistry, molecular mechanics and quantum chemistry.

The molecules of a pure chemical compound form generally an ensemble
of molecular individuals that differ in geometry and energy. Thus it is
generally impossible to represent a chemical compund adequately by the geometry of a rigid molecular model.

In modern stereochemistry it is often necessary to analyze molecular
relation within *ensembles* and *families* of stereoisomers and permutation
isomers, including molecules whose geometric features are changing with
time. Accordingly, there is definitely a need for new types of ideas,
concepts, theories and techniques that are usable beyond the scope of
customary methodology. This is why the present text was written.

The majority of organic molecules studied in modern stereochemistry
are flexible; depending on the observation conditions, they undergo a
variety of internal motions which are known to play a significant role in
determining their chemical behavior. There may be no chemically meaningful

rigid model that expresses the essential spatial features of such a molecule: for example, the fluxional motions of bullvalene [2] pass through arrangements that differ in chemical constitution, so that bullvalene cannot be meaningfully represented by any single, rigid model.

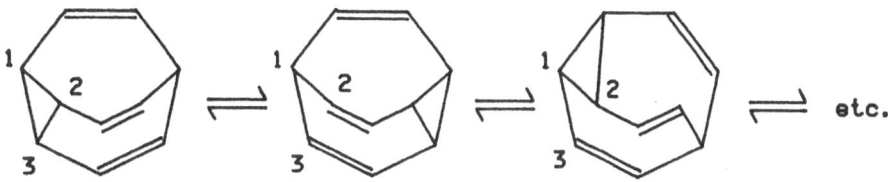

Moreover, even in cases where a flexible molecule can be described by a reasonable geometric model, the classical geometric considerations may not correctly predict the observed behaviour, as is illustrated by the mixed ester of (+) and (-)-menthol with 2,2',6,6'-tetra-nitro-4,4'-diphenic acid

which was observed by Mislow [3] to be achiral, despite the fact that it has no conceivable achiral conformation (see I,2).

Though various modifications and extensions of the classical geometry based stereochemical principles have been devised for treating individual flexible molecules [4], no single approach relying solely on geometry has been found to be universally applicable for describing the stereochemistry of such molecules which correspond to ensembles of interconverting conformations. Indeed, the proliferation of ad hoc treatments for individual molecules has generated ambiguities and misunderstandings (see I).

Thus, there is a growing insight [5,6] that new ideas beyond geometry are needed to cope with the great variety of rigid and nonrigid stereo-chemical systems. The purpose of this book is to provide a solid founda-tion for a completely general and rigorous unified treatment of stereo-chemistry. Though in special cases, our theory may be more cumbersome to use than the more familiar and less rigorous geometric, or energetic-geo-metrical methods, our theory is nevertheless applicable in all cases and can be used to determine the validity of conclusions reached by using either other models or simply ad hoc procedures.

In our view, stereochemistry cannot be treated adequately by con-sidering the molecular structure and the molecular chemistry separately. Rather, it is an interaction of these two features that is fundamental, and the basic concepts of stereochemistry should therefore reflect this inter-action. To develop this viewpoint, we first need a precise way, applicable to all molecules, for describing the molecular structure relevant to a stereochemical question, and the rearrangements of this structure that will be considered and/or allowed. We accomplish this by considering a given molecule to consist of a set of n sites (called the molecular skeleton) and a set of n ligands, and defining a structure to be a placement of the ligands on the sites. In this representation of molecular structure, the skeleton need not be rigid, nor even contiguous; the placement of a spec-ified ligand on a specified site is a well-defined operation, requiring only that the site be identifiable and not that it always be located at the same place in space. By permuting the ligands, we can then uniquely de-scribe all the possible rearrangements of the molecular structure, even for flexible molecules. The permutation isomers of a given molecule are obtained by placing the ligands on the sites in all the possible different ways [7].

Our fundamental notion is that of the chemical identity group. This is defined for molecules in which the ligands are all chemically distinguishable, and consists of all the ligand permutations that preserve the chemical identity of the given molecule. To describe its construction in heuristic terms, let us assume that we have made a film of a (possibly flexible) molecule, and that we have selected a "snapshot" as a reference. A permutation of the ligands on that reference is said to preserve the chemical identity of the molecule if the resulting molecule is geometrically identical to some frame in the film. The set of all such ligand permutations will form a group S, which we call the chemical identity group of the molecule.

Note that we do not deal only with symmetries and geometries, as do the customary permutational approaches to stereochemistry: we use permutations simply to express rearrangements of the ligands, and not a symmetry of some geometric system. Our chemical identity group expresses the geometry/chemistry interaction in a given molecule; its construction for a given molecule requires knowledge of a momentary spatial arrangement, the manner in which that arrangement changes under the given observation conditions, and the chemistry of the molecule. Although S is obtained by considering the "least symmetric" molecule of a skeletal class, it can be used to obtain the stereochemistry of molecules in that skeletal class having sets of chemically indistinguishable ligands.

The notions of chemical identity group and permutational isomerism provide the foundation for our formalization of stereochemistry. With this formalism, we can unambiguously express concepts relevant for stereochemistry, determine their interrelations, and obtain solutions to various stereochemical problems. For example, the number of chemically distinct isomers in a family of permutation isomers can be enumerated, and representatives of those isomers exhibited; chirality phenomena can be explained and predicted; the idea of an isomerization can be formulated precisely and the possible intermediates, or mechanisms, can be determined. Within a family of permutation isomers, the possible networks of isomerizations and their pathways, can be found.

Our approach does not involve energy considerations. It is mainly qualitative in the sense that we can describe all the possible outcomes of a given stereochemical experiment. In those cases where more than one outcome is indicated to be theoretically possible, our method serves to exclude possibilities that may initially appear to be attractive.

Hic liber omnis divisus est in partes tres. The purpose of the first part is to familiarize the reader with the main concepts and some of the essential principles. The second part contains the formal mathematical exposition of the theory and the third part consists of applications to current stereochemical problems.

As a more detailed description: In Chapter I, the main classical concepts are discussed, and their shortcomings when applied to flexible molecules are noted. Unambiguous definitions of stereoisomerism and chirality are then proposed which apply to all molecules, flexible or not. In Chapter II, the notions of permutational isomerism and chemical identity group are presented. The main results and some of the techniques are discussed in broad terms, with their use in studying chirality and isomerizations illustrated by applications to some simple stereochemical systems. In

Chapter III, it is shown that our general principles lead immediately to
the concept of the asymmetric carbon atom, and our description of its
structure is then contrasted with the classical description. Chapters IV-VI
are self-contained, presenting the mathematical details of the theory, with
the chemical motivation and interpretation of the mathematical results in
chemical terms always being given. A general treatment of conformational
flexibility is given in the last chapter of this section; it is shown that
the chemical identity group of such a molecule is the semidirect product of
subgroups having a clear-cut chemical meaning. The mathematics needed to
follow this development is elementary finite group theory; for the convenience of the reader and also to indicate the terminology used, a brief
account of the required group-theoretic background is given in an appendix.
Chapter VII contains various applications of the theory to: chirality and
hyperchirality [9]; mechanisms and modes; permutational isomerization such
as Berry pseudorotation and turnstile rotation; sigmatropic 1.5-hydrogen
shifts; conformation analysis; isomerization graphs; the bullvalene problem; the enumeration of isomers and the A.v.Baeyer - E.Fischer - J.van't
Hoff discussion concerning the stereoisomers of trihydroxyglutaric acid;
the S_N2 and related processes. In Chapter VIII a stereochemical nomenclature system based on permutational isomerism and the chemical identity
group is proposed.

One of the particular advantages of the theory of chemical identity
groups is that its computer assisted applications to stereochemistry use
only strikingly simple algorithms. For the solution of even extremely complicated stereochemical problems it suffices to have a computer program
capable of multiplying and conjugating permutations and of generating

cosets and double cosets from given permutations and subgroups of symmetric groups. Small computers from a TI 59 pocket calculator upwards suffice for the implementation of such programs [10], but comfortable programs require at least a small computer with the capabilities of an APPLE II.

References

[1] G. Wittig, "Stereochemie", Akademische Verlagsgesellschaft, Leipzig 1930; K. Freudenberg, "Stereochemie", Deuticke, Leipzig 1933; J. Weyer, Angew. Chem. 86, 604 (1974); Angew. Chem. Int. Ed. 12, 591 (1974).

[2] W. v. E. Doering and W. R. Roth, Angew. Chem. 75, 27 (1963); Angew. Chem. Int. Ed. Engl. 2, 24 (1963); G. Schröder, ibid., 75, 722 (1963); 2, 694 (1963); J. F. M. Oth, R. Merènyi, G. Engel and G. Schröder, Tet. Lett. 1966, 3377; J. F. M. Oth, R. Merènyi, H. Röttele and G. Schröder, Chem. Ber. 100, 3538 (1967).

[3] K. Mislow, Science 120, 232 (1954); Trans N. Y. Acad. Sci. 19, 298 (1957); K. Mislow and R. Bolstad, J. Amer. Chem. Soc. 77, 6712 (1955).

[4] see e. g.: H. C. Longuet-Higgins, Mol. Phys. 6, 445 (1963); J. E. Leonard, G. S. Hammond and H. E. Simmons, J. Amer. Chem. Soc. 97, 5052 (1975).

[5] J. Gasteiger, P. D. Gillespie, D. Marquarding and I. Ugi, Topics Curr. Chem. 48, 1 (1974); K. Mislow and P. Bickart, Isr. J. Chem. 15, 1 (1977); E. L. Eliel, ibid. 15, 7 (1977); R. G. Woolley, J. Amer. Chem. Soc. 100, 1073 (1978); A. T. Balaban, A. Chiriac, I. Motoc and Z. Simon, in: "Steric Fit in Quantitative Structure-Activity Relations", Lecture Note Series, Vol. 15, Springer, Heidelberg 1980; J. Dugundji, R. Kopp, D. Marquarding and I. Ugi, Topics Curr. Chem. 75, 165 (1978); K. Mislow and J. Siegel, J. Amer. Chem. Soc. (in press); see also: J. Brocas, M. Gielen and R. Willem, "The Permutatiuonal Approach to Dynamic Stereochemistry", McGraw-Hill, New York 1983.

[6] J. Dugundji, J. Showell, R. Kopp, D. Marquarding and I. Ugi, Isr. J. Chem. 20, 20 (1980).

[7] Permutational isomerism, the concept as well as its terminology, was introduced in 1970 [8].

[8] I. Ugi, D. Marquarding, H. Klusacek, G. Gokel and P. Gillespie, Angew.
Chem. 82, 741 (1970); Angew. Chem. Int. Ed. 9, 703 (1970).

[9] J. Dugundji, D. Marquarding and I. Ugi, Chem. Scripta 9, 74 (1976);
11, 17 (1977).

[10] Such a program has recently been implemented for a TI 59 pocket calculator by I. A. Ugi,jr.; personal computers e. g. the APPLE II suffice
generally for the required permutational computations.

CONTENTS

PART II.

The Mathematical Theory of the Chemical Identity Group

P A R T   III.

Application of the Theory of the Chemical Identity Group to Actual
Current Stereochemical Problems

PART I

THE PERMUTATIONAL APPROACH TO STEREOCHEMISTRY

C H A P T E R   I

THE DESCRIPTION OF MOLECULAR STRUCTURE

The classical description of molecules is based on their empirical formula, constitution, configuration, and conformation. In this chapter, we shall briefly review these concepts and also that of chirality, as they are commonly understood. A discussion of some structural features of flexible molecules indicates that purely geometrically based concepts of configuration, conformation, and chirality so well as some currently used modifications of those concepts cannot be universally applied in a consistent manner. We then propose a unified conceptual framework for stereochemistry, including new definitions of stereoisomerism and chirality. These definitions are based on chemistry, rather than on geometric models, and have an unambiguous meaning for all molecules.

1. The classical Description of rigid Molecules

The stereochemical structure of a chemical compound is considered to be known when the chemical constitution, configuration, and conformation of its molecules have been specified [1]. This information is generally obtained by various types of experiment, both physical and chemical, such as X-ray and neutron diffraction, NMR spectroscopy, structure elucidation by chemical methods as, for example, the enumeration and identification of stereoisomers, degradation, and stereoselectivity studies, etc.

As an introduction to the terminology and notation, we first review the customary concepts.

The empirical formula of a molecule indicates how many atoms of the various chemical elements this molecule contains. For example, the empirical formula for ammonia is NH_3, indicating that it consists of one nitrogen atom and three hydrogen atoms.

In molecules having the same empirical formula, the atoms may have different covalently connected neighbors, so we are led to the concept of chemical constitution. The chemical constitution of a molecule is specified by stating for each constituent atom its covalent bonds and the atoms to which it is connected by those bonds; frequently there is also a statement about the placement of "free" valence electrons [2,3]. This information is customarily conveyed by a constitutional formula, a diagram showing the interconnections of the atoms. The use of BE-matrices [2,3] to represent the chemical constitution of a molecular system is another method: The rows and columns of a BE-matrix are assigned to the atomic cores and its entries represent the distribution of the valence electrons[*]. For example, the chemical constitution for ammonia can be described by the customary constitutional formula 1 so well as by the BE-matrix B:

$$
\begin{array}{c}
\quad\quad\quad N\ H\ H\ H \\[4pt]
\begin{array}{c} N \\ H \\ H \\ H \end{array}
\left(
\begin{array}{cccc}
2 & 1 & 1 & 1 \\
1 & 0 & 0 & 0 \\
1 & 0 & 0 & 0 \\
1 & 0 & 0 & 0
\end{array}
\right)
\end{array}
$$

1   B

----

[*] The use of BE-matrices is particularly convenient for computer assisted documentation and for the manipulation of constitutional information about molecules. Moreover, the algebra of the BE- and R-matrices is the basis for a universal theory of constitutional chemistry [2] that is used to construct computer programs for the deductive solution of a variety of chemical problems [3].

Two molecules with the same empirical formula but different chemical constitution are called constitutional isomers. For example, n-butane **2** and
iso-butane **3** are constitutional isomers.

The fact that compounds whose molecules have the same chemical constitution may still be chemically distinguishable indicates that a more
detailed view of molecular structure is needed. Stereochemistry began with
the Le Bel - van't Hoff concept of the asymmetric carbon atom [4-6] (i. e.
the tetracovalent carbon atom having a valence skeleton with tetrahedral
symmetry and carrying a set of four different ligands) which showed that
the chemical behavior of a molecule may well be related to the spatial
arrangement of its atoms (see III).

In the standard terminology, molecules having the same chemical constitution but different spatial arrangements of their atoms are called
stereoisomers. They may differ configurationally and/or conformationally.

The configuration of a molecule with a monocentric skeleton is
described by the placement of ligands on the hypothetical valence state,
4 - 12, of its central atom as a skeleton [1,7].

sp^3 (C$_{3v}$)

7

sp^3 (T$_d$)

8

dsp^2 (D$_{4h}$)

9

dsp^2 (C$_{2v}$)

10

dsp^3 (D$_{3h}$)

11

d^{2}sp^3 (O$_h$)

12

Although these skeletons are not necessarily rigid under all observation conditions, the above representations have been quite effective in relating their structural and chemical properties [8].

Attempts to establish such relations for molecules with polycentric skeletons, by expressing their configuration in terms of monocentric sub-units, have been less successful [7]: no rules for specifying the sub-units uniquely have been formulated; and unstated assumptions in various studies, such as the requirement of independent free rotation of the sub-units, or the requirement that the selected sub-units should not be constitutionally equivalent, have led to some misunderstandings and confusion about what general principles are applicable in any specific situation [1]. Even in problems of enumerating the stereoisomers of such a molecule, it is not always clear which molecules, stereoisomers or permutation isomers, are to be counted [7] (see also VII).

Configurational isomers that are mirror images of each other are of particular interest. The concept of chirality was introduced by Lord Kelvin in 1892 [9]; he defined an object to be chiral if it is not superimposable onto its mirror image by rigid motions (i. e. rotations and translations). This definition, applied to the models of rigid molecules, gives the notion of chirality currently used in stereochemistry; a chiral molecule and its mirror image are called an enantiomer pair or, more simply, enantiomers. Since chirality was first noted in the case of chemical compounds whose molecules have asymmetric carbon atoms, asymmetry [8] has long been considered to be an essential characteristic of chemical chirality [10].

2. Nonrigid Molecules

In classical stereochemistry constitution and rigid configurations are the only molecular features considered to be chemically relevant. However, in recent decades, improved experimental techniques/measurements have disclosed an increasing number of nonrigid molecules, with many different types of flexibility [11]. In this section, we discuss some of the types of internal motions that flexible molecules can have, and assess some of the methods used to adapt the static notions of conformation/configuration to study the stereochemistry of flexible molecules.

In a vibrating ethene molecule 13, only slight changes of the interatomic distances and the bond angles take place at ambient temperatures; there is a high energy barrier opposing the relative rotation of the methylene units and thus no significant conformational changes occur.

13   14

In ethane **14**, however, there is a low energy barrier opposing relative
rotation of the methyl groups about the C–C bond; the bond angles remain
fairly constant. Under the customary observation conditions, the geometry
of an ethane molecule therefore changes rapidly with time [12]; at room
temperature, rotation about the carbon–carbon bond is approximately 10^9
revolutions/sec [13].

No such rigid bond angles are observed for nitrogen atoms; and in nitrogen compounds such as ammonia, the flexional vibrations have amplitudes
so large that the nitrogen atom seems to pass through the plane defined by
the three hydrogen atoms.

15a   15b

In tertiary amines **15** with three different ligands, this type of flexibility leads to configurational inversion (i. e. a configuration is converted into its enantiomer).

In the above types of intramolecular motions, the molecular geometry
changes without alteration of the constitution: no bonds are broken or
made. But in fluxional molecules such as bullvalene (see Introduction), the
internal changes also involve the making and breaking of covalent bonds and
changes in covalently bound neighbours; these are described as Cope rearrangements (see VII,5). With suitable substituents, the fluxional changes
in bullvalene pass through stages having different chemical constitutions:
the coordination number of some atoms may change (for example, a carbon
atom surrounded by three carbon and one hydrogen atom in one species
becomes bound to two carbon and one hydrogen atom in another species) [14].

Thus, the nonrigidity of molecules can be of various types, so that the geometries of the molecules belonging to a given pure chemical compound can differ noticeably from each other, and may be changing considerably in time. However, the classical stereochemical concepts are expressed in terms of fixed geometries and rigid models of molecules. In order to apply the classical ideas to these new cases, some method for assigning a fixed geometry to a nonrigid molecule is therefore needed. We describe three (of the many) different methods that are used, and indicate their limitations.

One approach relies on time-average geometries [15]: the molecules are described geometrically by a rigid model, with the intramolecular motions treated as deviations from time-average positions. For ammonia, this is an arrangement in which all the atoms and bonds are coplanar. With this simple trick, the stereochemistry of ammonia and its suitably substituted derivatives can be treated in the classical way: for example, this representation accounts for the fact that no enantiomers are obtained in the case $NL_1L_2L_3$, so long as the observation conditions assure that inversion of the nitrogen skeleton is rapid. The concept of a time-average geometry is also useful for the prediction and interpretation of the presence or absence of chirality in many complex flexible molecules, such as the substituted cyclohexanes [15,16]. However, the time-average geometry approach cannot be universally applied: for example, no such unique geometry is available for ethane or for bullvalene; indeed, even when such a geometry exists and is useful, it may not represent any possible momentary arrangement of the molecule (as in cyclohexane models having a planar carbon ring).

A second approach uses energy considerations to develop a geometric model, representing the molecule by a conformation that has the lowest molecular energy; the intuitive justification is that the more stable (i. e. less energetic) conformation will be present in higher concentration [17,18]. In the case of ethane, the staggered conformation is the more stable, being favored by 13 kJoule/mol over the eclipsed one. The energy difference has been variously ascribed to the H-H repulsions of the eclipsed hydrogens (although the distance between them seems to be too great to generate any significant interaction), to the nature of the C-C bond, and to non-bonded interactions [19]. This approach to getting a reasonable rigid model for a flexible molecule is satisfactory for many of the stereochemical problems that are of interest to the organic chemist. But again this approach is not universally applicable, because there is frequently no single unique conformation with lowest molecular energy; for example, in "dimethyl-polyacetylenes" 16 with a sufficiently large n, the

16

molecular energy differences between any of its infinitely many conformations (staggered, eclipsed and intermediate) will probably be so small that they are not physically measurable.

A third approach treats a nonrigid molecule as an ensemble of interconverting molecules, and applies the methods of quantum chemistry and statistical thermodynamics to derive an equilibrium position [20]. This has been successful in many cases to determine, say, the most prevalent geometric arrangement; it requires great precision in calculations, since small errors (e. g. 25 kJoule/mol) in free energy calculations can cause large differences in predicted results (such calculations become extremely complicated for large molecules). Again, this method is not applicable to all stereochemical problems. For example, it is not adequate for describing interconversion mechanisms, nor has it been too successful in describing chirality related phenomena.

The treatment of chirality for flexible molecules seems to be beyond the scope of geometric considerations: in view of the non-existence of unique time-average geometries, or suitably describable thermodynamic equilibria, the meaning of the classical notion of mirror image becomes obscure for flexible molecules. One approach has been to declare a flexible molecule achiral if at least one of its conformations is achiral. There are, however, achiral molecules with no achiral conformation. For example (see Introduction), Mislow [21] demonstrated that the mixed ester of (+) and (-)-menthol with 2,2',6,6'-tetra-nitro-4,4'-diphenic acid **17** is chemically achiral [10]; however, the molecule **17** has no conceivable geometrically achiral conformation. It is converted into its enantiomer $\overline{17}$ by a feasible 90° internal rotation of its diphenyl moiety.

The above discussion indicates that the classical notion of molecular geometry is of limited use for understanding the stereochemistry of the organic compounds that arise in modern chemistry [7, 10]: appropriate rigid models cannot always be found and, even if such a model exists and is useful, it still may not represent any physical reality. These considerations have generated a growing realization that new ideas, going beyond geometry,

17

$\overline{17}$

are needed in order to develop a logically consistent treatment of stereo-chemistry [11]. This realization is reinforced by considering Cram's and Prelog's rules [22–24], which are among the best-known examples that show how plausible, and yet fallacious, the purely geometric visualization of stereochemical processes can be.

These rules, which describe the preferred stereochemical outcome of certain asymmetric syntheses, are given by mnemotechnical diagrams such as

18   19

Cram's rule (steric bulk R_L > R_M > R_S)

and

20   21

Prelog's rule.

The authors have clearly stated that they intend these diagrams to express
that the preferred product of the reaction is formed as if the molecules
of the reacting carbonyl compound existed mainly in the indicated conformation and were attacked from their sterically less hindered sides.
Nevertheless, these and other such rules have often been misunderstood and
quoted in a misleading manner to suggest that the above geometric pictures
give a real explanation of the preferred stereochemical course of the
stereoselective reactions under consideration. However, a later critical
study of asymmetric syntheses has led to the conclusion that geometric

models of the reactants are useless and confusing in the treatment of stereoselective reactions of nonrigid molecules [18,25,26], and that a combination of group theory and statistical thermodynamics is needed for the interpretation of such reactions [26].

3. Definition of Chemical Identity, Stereoisomerism, and Chemical Chirality

Most organic molecules are not rigid; moreover, as we have just seen, many of them cannot be assigned a uniquely determined geometry that can adequately express all their stereochemical features. Thus, it is desirable that the basic concepts of stereochemistry be expressed, so far as possible, in terms independent of individual molecular geometries; the concepts should involve only those molecular features that are chemically relevant and have unambiguous meaning for all molecules, rigid or flexible. To standardize the terminology, we shall give definitions that meet these criteria and that we will use throughout this book.

The most inclusive definition of chemical identity is strictly empirical and expresses the impossibility of separating identical molecules by chemical means. Based on this general notion, we make the

<u>3.1 Definition</u> Under given observation conditions, two molecules are called chemically identical if they belong to the same chemical compound.

This implies that two molecules will be chemically identical whenever there exists some spatial arrangement of the atoms that both molecules can acquire by means of the rotations, translations, and intramolecular motions possible under the given observation conditions.

3.2 <u>Definition</u>   Two molecules are called stereoisomers if they have the same chemical constitution, but are not chemically identical[*]).

Since constitution involves only the interconnections of the atoms of a molecule, and not their relative spatial positions, this definition is unambiguous for all molecules, flexible or not (see I,1). Note that stereo-isomerism, which was previously expressed in geometric terms, or by using the often vague notions of configuration and/or conformation, is now ex-pressed by a more precise concept.

3.3 <u>Definition</u>.   Two chemical compounds are stereoisomeric if their mole-cules are stereoisomers.

It is interesting to note that this definition avoids all reference to individual molecular geometries and that it records the commonly accepted meaning of the term as it is used by most chemists.

Chirality plays a major role in the description and classification of stereoisomers [27]. Chirality as defined by Lord Kelvin [9], and as it is used as a stereochemical concept, is purely geometric, being applicable only to molecules with rigid skeletons and idealized ligands. Since molecu-les are generally not rigid, a different concept of chirality is needed for stereochemistry. The following definition applies to all molecules, flexi-

----

[*] This differs somewhat from the customary definitions of stereoisomers [8].

ble or not, and agrees with the classical concept whenever the molecule is rigid:

3.4 Definition [10].   Under given observation conditions, a molecule is chemically achiral, if each momentary geometry of the molecule can be superimposed on its mirror image (the geometric enantiomer) by using only the rotations, translations, and intramolecular motions that can occur under the given observation conditions.   A molecule that is not chemically achiral is called chemically chiral.

From this definition, an amine with three chemically distinct ligands 15 is chemically achiral, as is meso-tartaric acid 22 (22a is interconverted with its enantiomer $\overline{22a}$ and the achiral conformation 22b through internal rotation about the C-C σ-bond) [10].

22a   22a

22b

Mislow's ester 17 is also chemically achiral, even though it has no geometrically achiral conformation: the definition implies that the existence
of an achiral geometric arrangement is not necessary for chemical
achirality. Observe that a molecule is chemically chiral under given
observation conditions, if it has at least one geometric arrangement that
cannot be superimposed on its mirror image by the rotations, translations,
and intramolecular motions that can occur under the given conditions [10].

References

[1] J. Gasteiger, P. D. Gillespie, D. Marquarding and I. Ugi, Top. Curr.
    Chem. 48, 1 (1974).
[2] J. Dugundji and I. Ugi, Top. Curr. Chem. 39, 19 (1973).
[3] I. Ugi, J. Bauer, J. Brandt, F. Friedrich, J. Gasteiger, C. Jochum and
    W. Schubert, Angew. Chem. 91, 99 (1979), Angew. Chem. Int. Ed. 18, 111
    (1979); I. Ugi, J. Bauer, J. Brandt, J. Friedrich, J. Gasteiger, C.
    Jochum, W. Schubert and J. Dugundji, in: "Computational Methods in
    Chemistry", ed.: J. Bargon, Plenum Press, New York, N. Y. 1980, p.
    275; I. Ugi, J. Bauer, J. Brandt, J. Dugundji, R. Frank, J. Friedrich,
    A. v. Scholley and W. Schubert, in: "Data Processing in Chemistry",
    ed.: Z. Hippe, PWN-Polish Sc-Pub. Warsaw 1981, p. 219; J. Brandt, J.
    Bauer, R. M. Frank and A. v. Scholley, Chem. Scripta 18, 53 (1981); C.
    Jochum, J. Gasteiger, I. Ugi and J. Dugundji, Z. Naturforsch. 37B,
    1205 (1982); J. Bauer and I. Ugi, J. Chem. Res. 1982(S) 298, (M)310,
    3201.
[4] J. H. van't Hoff, Arch. Neer. Sci. Exactes Natur. 9, 445 (1874); Bull.
    Soc. Chim. France [2] 23, 295 (1875); "The Arrangement of Atoms in
    Space", Longmans Green, London 1898.
[5] J. A. LeBel, Bull. Soc. Chim. France [2] 22, 337 (1874).
[6] J. Weyer, Angew. Chem. 86, 604 (1974); Angew. Chem. Int. Ed. 12, 591
    (1974); J. K. O'Loane, Chem. Rev. 80, 41 (1980).
[7] I. Ugi, D. Marquarding, H. Klusacek, G. Gokel and P. Gillespie, Angew.

Chem. <u>82</u>, 741 (1970); Angew. Chem. Int. Ed. <u>9</u>, 703 (1970).

[8] E. L. Eliel, "Stereochemistry of Carbon Compounds", McGraw-Hill, New
     York 1962; K. Mislow, "Introduction to Stereochemistry", Benjamin, New
     York 1965; J. F. Stoddart, in "Comprehensive Organic Chemistry", Vol.
     1, eds.: Sir D. H. R. Barton and W. D. Ollis, Pergamon Press, Oxford
     1979, p. 13; see also: W. Bähr and H. Theobald, "Organische
     Stereochemie-Begriffe und Definitionen", Springer-Verlag, Heidelberg,
     1973, p. 107; B. Testa, "Principles of Organic Stereochemistry",
     Marcel Dekker, New York 1979; idem, "Grundlagen der Organischen
     Stereochemie", Verlag Chemie, Weinheim 1983.

[9] Lord Kelvin, "Baltimore Lectures", Clay and Sons, London 1904, p. 436,
     619.

[10] J. Dugundji, R. Kopp, D. Marquarding and I. Ugi, Top. Curr. Chem. <u>75</u>,
     165 (1978).

[11] K. Mislow and P. Bickart, Isr. J. Chem. <u>15</u>, 1 (1977); E. L. Eliel,
     ibid. <u>15</u>, 7 (1977); R. G. Woolley, J. Amer. Chem. Soc. <u>100</u>, 1073
     (1978).

[12] E. L. Eliel, N. L. Allinger, S. J. Angyal and G. A. Morrison, "Con-
     formational Analysis", Interscience, New York 1965; M. Hanack, "Con-
     formation Theory", Academic Press, New York 1965; G. Chiurdoglu, ed.,
     "Conformational Analysis", New York 1971; J. Dale, "Stereochemie und
     Konformationsanalyse", Verlag Chemie, Weinheim 1979.

[13] C. E. Wintner, "Strands of Organic Chemistry", Holden-Day, San
     Francisco 1979, p. 9.

[14] W. v. E. Doering and W. R. Roth, Angew. Chem. <u>75</u>, 27 (1963); Angew.
     Chem. Int. Ed. Engl. <u>2</u>, 24 (1963); G. Schröder, ibid. <u>75</u>, 722 (1963);
     <u>2</u>, 694 (1963); J. F. M. Oth, R. Merènyi, G. Engel and G. Schröder,
     Tet. Lett. <u>1966</u>, 3377; J. F. M. Oth, R. Merènyi, H. Röttele and G.
     Schröder, Chem. Ber. <u>100</u>, 3538 (1967).

[15] J. E. Leonard, G. S. Hammond and H. E. Simmons, J. Amer. Chem. Soc.
     <u>97</u>, 5052 (1975); see also: ref. [7,16].

[16] P. Gillespie, P. Hoffmann, H. Klusacek, D. Marquarding, S. Pfohl, F.
     Ramirez, E. A. Tsolis and I. Ugi, Angew. Chem. <u>83</u>, 691 (1971); Angew.
     Chem. Int. Ed. <u>10</u>, 687 (1971).

[17] V. Prelog, in: "Perspectives in Organic Chemistry", ed.: Sir A. Todd,
     Interscience, New York 1956.

[18] I. Ugi, in: "Jahrbuch 1964 der Akademie der Wissenschaften in
Göttingen", Vandenhoek & Rupprecht, Göttingen 1965, p. 21.

[19] A. Streitwieser and C. H. Heathcock, "Introduction to Organic
Chemistry", Mc Millan, New York 1981, p. 83.

[20] see e.g.: N. Trong Anh and O. Eisenstein, Tet. Lett. 1976, 155.

[21] K. Mislow, Science 120, 232 (1954); Trans N. Y. Acad. Sci. 19, 298
(1957).

[22] D. J. Cram and F. A. Abd Elhafez, J. Amer. Chem. Soc. 74, 5828, 5851
(1952); see also: D. Kruger, A. E. Sophitz and C. A. Kingsbury, J.
Org. Chem. 49, 778 (1984); J. Mulzer, Nachr. Chem. Tech. 32, 17 (1984)
and references therein.

[23] V. Prelog, Helv. Chim. Acta 36, 308 (1953).

[24] J. D. Morrison and H. S. Mosher, "Asymmetric Organic Reactions",
Prentice-Hall, Englewood Cliffs, N.J. 1971.

[25] I. Ugi, Z. Naturforsch. 20B, 405 (1965).

[26] E. Ruch and I. Ugi, Theoret. Chim. Acta (Berl.) 4, 287 (1966); Top.
Stereochem. 4, 99 (1969).

[27] V. Prelog (Nobel Lecture), Science 193, 17 (1976); see also: E. F.
Meyer, J. Comput. Chem. 1, 229 (1980); V. Prelog and G. Helmchen,
Angew. Chem. 94, 614 (1982); Angew. Chem. Int Ed. 21, 567 (1982).

THE CHEMICAL IDENTITY GROUP

In this chapter, we describe the construction of the chemical identity group [1] as well as the racemate group for a given compound, and present some of the techniques that will be used in this book to represent and interpret stereochemistry.

1. *Families of Permutation Isomers*

In order to describe all the stereochemical changes that a given molecule can undergo under given observation conditions, it is necessary to specify the parts that can be rearranged, the changes that are permitted in the experiment under consideration, and to have some method for specifying those changes precisely.

We can accomplish all this by regarding the molecule to consist of a skeleton and a set of ligands, where we call ligands those atoms, or polyatomic groups, that can be permuted, and we call skeletal sites that part of the molecule which remains after all the ligands are removed [1-4]. To give the broadest scope to our considerations, we take the permissible molecular rearrangements to be all the distinct ways of placing the ligands on the skeletal sites. Any two molecules obtained by a ligand rearrangement are called permutationally isomeric, and the set of all the molecules obtained in this way is called a family of permutation isomers.

An exact description of all these molecules can be gotten by selecting one of them as the reference isomer X. We then choose one molecular individual E from the isomer X as a characteristic model, the reference model. Then any rearrangement of the given ligands on the skeletal sites is completely described by a permutation of the ligands on the reference model. For example, with the reference model E

$$3 \underset{5}{\overset{1}{\vert}} \overset{2}{\underset{4}{}} \quad \xrightarrow{(124)} \quad 3 \underset{5}{\overset{4}{\vert}} \overset{1}{\underset{2}{}} \quad \xrightarrow{(25)} \quad 3 \underset{2}{\overset{4}{\vert}} \overset{1}{\underset{5}{}}$$

$$E \qquad\qquad (124)E \qquad\qquad (25)(124)E$$
$$= (2541)E$$

the permutation (1 → 2 → 4 → 1) of ligands, which we write as (124) in the standard permutational notation, results in the molecule denoted by (124)E, and the permutation (25) performed on the ligands of the latter molecule gives (25)(124)E - which can be obtained by applying the product (25)(124)= (2541) of the permutations directly to E. Similarly, the molecule with the ligand placement

$$4 \underset{1}{\overset{3}{\vert}} \overset{5}{\underset{2}{}}$$

is simply (15243)E.

Thus, working within a family of permutationally isomeric molecules, we are able to unambiguously and precisely describe all the molecules that will be considered and compared in the given experiment [5].

The concept of permutational isomerism must be carefully distinguished from that of stereoisomerism. There exist permutation isomers which are not stereoisomers, and there are stereoisomers that are not permutation isomers [5]. Permutation isomers with a monocentric skeleton are always stereoisomeric, but in the case of permutation isomers with a polycentric skeleton

some members of a family of permutation isomers have the same chemical constitution and thus are stereoisomers, while others are only constitutional
isomers. This is illustrated by the following examples:

(a) (R)-alanine 2 and (S)-alanine $\bar{2}$ are stereoisomers (enantiomers) and at
the same time permutation isomers.

$$NH_2 \qquad\qquad CO_2H$$

$$CH_3 \quad C \quad H \qquad\qquad CH_3 \quad C \quad H$$

$$CO_2H \qquad\qquad NH_2$$

$$2 \qquad\qquad\qquad \bar{2}$$

(b) The rigid model 3 of dimethylamino-tetrafluorphosphorane is stereoisomeric to rigid model 4 of the transition state of its Berry pseudorotation
[6], but 3 and 4 are not members of the same family of permutation isomers.

$$(CH_3)_2N—P \qquad\qquad (CH_3)_2N—P$$

3   4

[7]
(D_{3h}-skeleton)   (C_{4v}-skeleton)

(c) The truxinic and truxillic acids 5 and 6 are permutation isomers with
a cyclobutane skeleton. However, they differ constitutionally and are thus
not stereoisomeric.

$$Ph \qquad\qquad\qquad\qquad COOH$$

$$Ph \qquad COOH \qquad\qquad Ph \qquad Ph$$

$$COOH \qquad\qquad\qquad COOH$$

5   6

For a molecule X with a given set of ligands the following Venn dia-

gram illustrates the relations between the various types of its isomers.

Let set A contain the permutation isomers and set B the stereoisomers of X.

Both are subsets of a larger set C which contains the isomers of X having

the same empirical formula as X. The intersection A ∩ B of A and B con-

sists of those permutation isomers of X which have the same constitution as

X, i. e. its stereoisomeric permutation isomers.

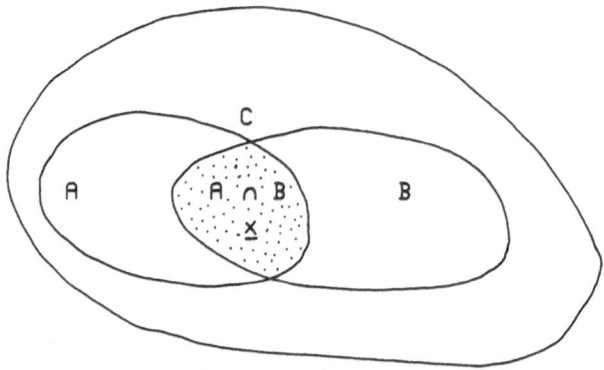

Many of the misunderstandings and misinterpretations in the stereo-

chemical literature are due to the fact that no distinction was made bet-

ween stereoisomers in general and those stereoisomers that are at the same

time permutation isomers.

Use of a permutational approach for the study of molecules was initi-

ated by Polya [9] in 1936, in his enumeration of certain types of isomers.

Polya's counting procedure was significantly extended by de Bruijn [10],

and modified by Ruch et al. [11]. The general concept and term permutatio-

nal isomerism was first explicitly defined in 1970 [5]; its distinction

from stereoisomerism was then also pointed out, and the representation of

permutation isomers by (2,n)-matrices and by their permutational descrip-

tors was   introduced. Subsequently Klemperer [12] and Nourse [13] published interesting contributions to permutation isomerism. The proceedings of a conference on the use of permutations in chemistry and physics [14] provide a survey of the recent literature of this field.

The traditional uses of permutation groups in stereochemistry [5] [9-15] have been successful in the solution of various stereochemical problems. However, in those approaches a given permutation can represent either a ligand exchange, or a possibly nonexistent skeletal symmetry operation, or an intraskeletal motion, which, as can be expected, generates conceptual difficulties. No universally applicable unified theory of stereochemistry has evolved from those studies, because they are essentially based only on elementary geometry, and we have seen in Chapter I that this foundation has inherent limitations.

2. *The Chemical Identity Group*

Let X be a given compound, which we assume to be pure and uniform; though the individual molecules of X can differ greatly in shape, they are, by definition I,3.1, all chemically identical.

Let us select one molecular individual from compound X as a characteristic model and define its skeleton and ligands (appropriate for the given experiment). This characteristic model is called the reference model, E. We shall assume that E has n ligands and, to simplify the exposition, that these ligands are all chemically distinguishable from one another[20].

Each permutation of the ligands of E gives a molecule representing some chemical compound, not necessarily X; the chemical identity group of X is determined by those permutations of the ligands of E that do in fact represent a molecule of a chemical compound identical to X. Note that we are relying on the unambiguous definition I,3.1, rather than on any purely geometric concept such as skeletal symmetry; in fact, for molecules with chemically distinct ligands the ligand interactions always cause the skeletons to deviate from their idealized symmetries. Accordingly, a chemical identity preserving ligand permutation does not necessarily bring the skeleton into self-coincidence, as is required in the conventional representation of skeletal point group symmetries and dynamic symmetries by permutations of skeletal sites or idealized ligands (see III,3).

To define the chemical identity group formally, recall that, if L is the set of ligands, then the set of all permutations of L, with the usual composition of permutations, forms a group SymL called the symmetric group on |L| objects (see Appendix). Using this terminology we can state precisely:

2.1 Definition. Let X be a given compound and E a reference model for X having a set L of chemically distinguishable ligands. Let S_X be the set of all permutations of the ligands of E that yield models chemically identical with E, all representing X. For reasons based entirely on the nature of chemistry and explained in Chapter IV, the set S_X will be a subgroup of SymL. We call S_X the chemical identity group of X.

The chemical identity group is the conceptual basis for our representation of the stereochemistry of molecules, flexible or not. We remark that the choice of any other reference model E' from X yields (cf. Chapter IV) the same group S_X.

As is well known, groups are frequently used to express geometric sym-metries [8]. Our chemical identity group is, however, new in concept and in intent. The chemical identity group does not express geometric symmetries so much as it expresses stereochemical realities. Moreover, even our index-ing system [1] is different from those used previously: our approach is based entirely on permutations of indexed ligands on a fixed model, and we do not assign indices to the skeletal sites at all (see e. g. ref. [11]).

If $|L| = n$, the ligands on E can be labeled $1, 2, ..., n$ in any way; different labelings will simply give conjugate (therefore isomorphic) sub-groups of the symmetric group S_n as the chemical identity group of X, a matter that does not affect the development of the theory. However, in order to standardize notation, as is necessary e. g. for documentation pur-poses (see VIII), indexing of the ligands is, in practice, determined by their chemical nature and is performed according to the CIP rules [16] or the CANON algorithm [17, 18] (see VIII, 5).

We will show later (IV, 2.5) that enumeration of the chemically dis-tinct permutation isomers of X is immediate, once its chemical identity group S_X is known: indeed, we have

2.2 **Proposition.** If all the ligands are chemically distinct, then X has exactly $|S_n|/|S_X|$ chemically distinct permutation isomers; and in fact, all the permutations belonging to a given left coset λS_X of S_X in S_n will generate the same isomer from E. Moreover, the chemical identity groups of the permutation isomers all belong to the conjugacy class of S_X in S_n.

If the ligands are not all chemically distinguishable, then the number

of the chemically distinct permutation isomers of X can be determined by an

additional straightforward algebraic procedure (see V,2).

We give examples to illustrate the construction of a chemical identity

group; note that information about both the chemistry and the geometry of X

is needed in order to construct its chemical identity group.

<u>2.3</u> <u>Example</u>. A phosphine derivative with three distinguishable ligands

can be represented by the model

E

We shall determine the chemical identity group of X under the assumption

that the chemical data suggests a rigid skeleton. Since there are three

ligands, we work in S_3; and for the reader's convenience, the group S_3

and its multiplication Table 1 is given in

second ⟍ first	e	(123)	(132)	(12)	(13)	(23)
e	e	(123)	(132)	(12)	(13)	(23)
(123)	(123)	(132)	e	(23)	(12)	(13)
(132)	(132)	e	(123)	(13)	(23)	(12)
(12)	(12)	(13)	(23)	e	(123)	(132)
(13)	(13)	(23)	(12)	(132)	e	(123)
(23)	(23)	(12)	(13)	(123)	(132)	e

We consider the chemical effect of each member of S_3 on E. First, (123)E is simply the molecule E rotated by 120°:

$$3 \diagup \overset{P}{\underset{2}{\big|}} {}_{\cdots 1} \quad \overset{(123)}{\longrightarrow} \quad 2 \diagup \overset{P}{\underset{1}{\big|}} {}_{\cdots 3}$$

E   (123) E

being rigid, (123)E is superimposable on E, so it is chemically identical to E, and therefore the permutation (123) belongs to S_X. The same is true for (132)E, so (132) $\in S_X$, and obviously e $\in S_X$. For the remaining permutations, no one of the molecules

$$3 \diagup \overset{P}{\underset{1}{\big|}} {}_{\cdots 2} \qquad 1 \diagup \overset{P}{\underset{2}{\big|}} {}_{\cdots 3} \qquad 2 \diagup \overset{P}{\underset{3}{\big|}} {}_{\cdots 1}$$

(12) E   (13)E   (23)E

is superimposable on E because the skeleton is rigid, so they are all chemically distinct from E. Thus, the ligand permutations that preserve the chemical identity of X are $S_X = \{e,(123),(132)\}$ and, as a glance at the multiplication table shows, S_X is indeed a group. Since $|S_3|/|S_X| = 6/3 = 2$, there are two chemically distinct permutation isomers, and they are represented by the distinct left cosets of S_X. The left coset of S_X different from S_X is (12)S_X and, from the table, (12)·$\{e,(123),(132)\}$ = $\{(12),(13),(23)\}$. These are the permutations changing E to the molecules listed above; and since these permutations belong to the same coset, the 2.2 proposition assures that these molecules are all chemically identical - which can be seen directly since they are superimposable on one another by rotations.

    Thus, only two chemically distinct isomers can be formed by rearranging the ligands on E, and we have produced models representing each one of these two isomers. Observing that (12)E and E are in fact enantiomers, we conclude that X is chiral.

2.4 Example. Consider the same skeleton as in the 2.3 example, but as encountered in a tertiary amine. The chemical information is that there exists only one compound, and no chemically distinct isomer can be formed. This means that all the different rearrangements of the ligands on the sites yield chemically identical molecules, so we conclude that now $S_X = S_3$. Note the importance of the chemistry in determining the chemical identity group: unlike the previous example, the role of the geometry is irrelevant in this case.

2.5 Example. Let the tetrasubstituted ethene derivative 7, compound X, be represented by the model E = 7a with a rigid skeleton.

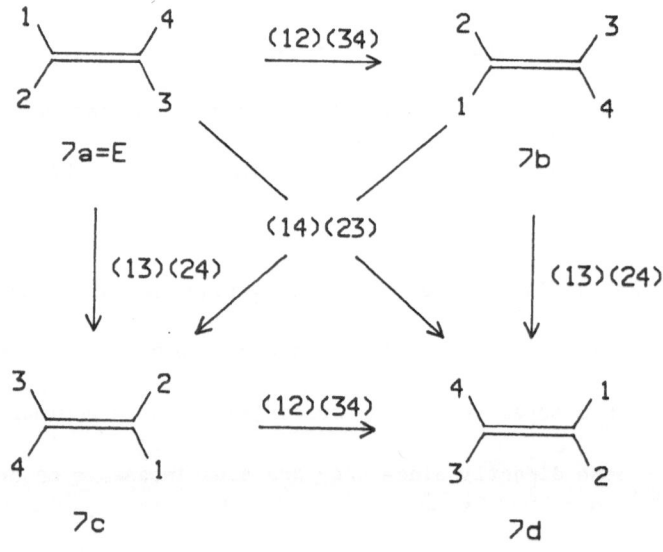

Since E has four ligands, S_7 is a subgroup of S_4, the symmetric group on four symbols.

Reference model 7a is converted into 7b by a 180° rotation about the axis of the C-C bond. This rotation changes $1 \rightarrow 2$, $2 \rightarrow 1$, $3 \rightarrow 4$, $4 \rightarrow 3$ which is signified by the permutation (12)(34). The two other such rotations are (13)(24) and (14)(23). Thus we have the chemical identity group $S_7 = \{e,(12)(34),(13)(24),(14)(23)\}$.

The group S_7 has four elements, i. e. its order $|S_7| = 4$, while the order $|S_4| = 4! = 24$. Using 2.2 proposition, the family of 7 consists of $|S_4|:|S_7| = 24:4 = 6$ permutation isomers.

2.6 **Example**. Consider an ammonium ion $L_1L_2L_3NH^+$ formed by protonation of a tertiary amine bearing three distinct ligands. In the presence of a trace amount of the amine the ammonium ion undergoes rapid proton transfer followed by inversion at the nitrogen, and only one compound can be distinguished chemically. In other words, all the different placements of the four ligands yield chemically identical molecules. This means that $S_X = S_4$ (see I,2 and III). Note again the importance of the chemistry in determining the chemical identity group.

If, however, the above ammonium ion existed in an environment that is sufficiently acidic to prohibit deprotonation of $L_1L_2L_3NH^+$ and reprotonation of $L_1L_2L_3N$, then the ammonium ion would not interconvert with its enantiomer, and its chemical identity group would be the alternating group A_4, i. e. all the even permutations in S_4 (see III,1).

3. Role of the Chemical Identity Group in Stereochemistry

We have seen that knowledge of both the chemistry C and the geometry G
of X uniquely determines the chemical identity group S_X of X. We can
express this symbolically as $S_X = F(C,G)$, i. e. S_X is a function of both C
and G.

However, the converse of this statement is not true: knowing S_X alone
permits only the determination of pairs (C,G) that are compatible with the
given S_X. Indeed, even knowledge of S_X and only one of C,G does not
determine the other uniquely.

Nevertheless the symbolic equation $S_X = F(C,G)$ indicates that S_X can
always be used to help express stereochemical facts algebraically in a
chemically consistent fashion. Assume, for example, that only C is known
and that there is insufficient information about the exact chemical archi-
tecture G of X. Then

(a) It may be possible to determine S_X from a sufficiently large set of
known or assumed data, and then find a compatible G (as we will see,
this is essentially the approach used by Le Bel - van't Hoff in their
determination of the geometry of an asymmetric carbon atom (see III);
and it is also an approach in the determination of isomerization
mechanisms).

(b) It may be appropriate to propose G, and to test the validity of that
proposal by finding whether or not the resulting $S_X = F(C,G)$ gives
results in agreement with experiment (e. g. the correct number of
distinct permutation isomers).

To illustrate this, we determined S_X of 8 on the basis of an assumed chemistry C (i. e. that there is only one permutation isomer) and a skeletal geometry G. The same S_X and C are compatible with a G in which the molecular skeleton has a D_{3h} symmetry (i. e. three two-fold rotational symmetry axes that lie in a single plane, a three-fold rotational axis perpendicular to that plane, etc. [7,8]) which is represented by a planar molecular structure 8.

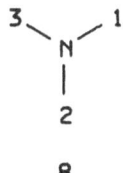

8

Nor is this the only geometry compatible with the S_X and C: for example, we can assume that under given observation conditions the molecule is vibrating so rapidly that the six arrangements 8d-8i belong to the same chemical compound; it follows then that $S_X = S_3$.

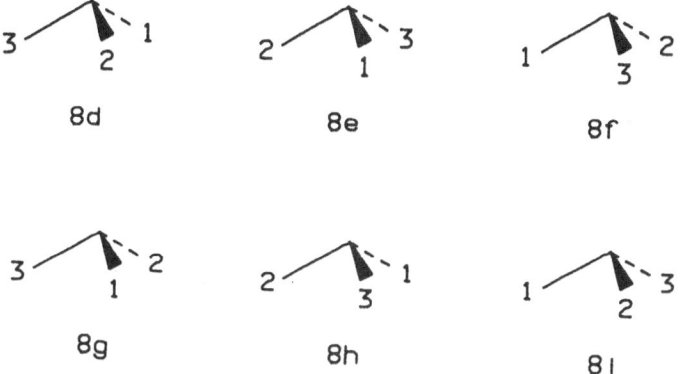

Note that because S_X contains two types of permutations, some chemically based and others geometrically based, the more permutations we ascribe to the geometry of the skeleton (i. e. the more symmetric we assume the skeleton to be) the fewer ligand permutations are ascribed to "chemical" reasons (see II,4, Table 2).

4. *Racemate Group*

The racemate group, which is an extension of the chemical identity group, is the fundamental tool in our study of chirality in families of permutation isomers (see VII,1).

Under given observation conditions, we say that X has an enantiomer $\overline{X}$, if each geometric arrangement of a molecule from X is the mirror image of some molecule from $\overline{X}$ and conversely. A mixture of equal numbers of molecules from X and $\overline{X}$ is called the racemate of X.

Let X be the reference isomer of a family of permutation isomers with all ligands chemically distinguishable. If X has an enantiomer $\overline{X}$ belonging to the same family of permutation isomers, then X and $\overline{X}$ have the same skeleton; we then say that the skeleton of X is achiral. In this case, any permutation of the ligands of X that preserves the chemical identity of X also preserves the chemical identity of $\overline{X}$, so that both X and $\overline{X}$ have the same chemical identity group. The chirality of such isomers must then be due to differences in the placement of the ligands on the skeleton.

4.1 <u>Definition</u>. The union R_X of the set of all permutations that preserve the chemical identity of X and the set of all permutations that interconvert X with its enantiomer $\overline{X}$, is in fact a group, called the racemate group of X (see IV,5).

It turns out that $S_X \subset R_X$ is a subgroup of order 2 (see V,5), therefore normal, in R_X. The coset $\overline{S}_X$ of S_X in R_X is called the enantiomer coset; any permutation belonging to the enantiomer coset will interchange X and $\overline{X}$ with each other.

4.2 **Example**. Consider a racemic mixture of the enantiomeric tertiary

phosphines X and X̄ (see 2.3 Example). The racemate group of X is S_3.

We will see that two chemically distinct permutation isomers (1) can

have the same chemical identity groups, but different racemate groups and

(2) can have different chemical identity groups, but the same racemate

groups (see VII,1.3). More importantly, (3) there are sets of chemically

distinct permutation isomers all having the same chemical identity group

and the same racemate group; we call such a set of isomers a hyperchiral

family [1]. Precise conditions under which (1), (2) and (3) can occur will

be given in IV,5.

The racemate group, which has been defined only when all the ligands

of the molecule are distinguishable, can also be used to determine if that

molecule with some subsets of its ligands made indistinguishable, is chiral

or not. Among the consequences of this, we can treat the vague concept of

prochirality in a consistent manner, not as a property inherent in a mol-

ecule, but as a property that a molecule has with respect to a specified

operation π, (such as permuting, substituting, adding, or removing lig-

ands). Thus, we will say X is prochiral with respect to the operation π if

the resulting molecule π(X) is chiral.

In the following table, we list the chemical identity and the racemate

groups for some molecules having four chemically distinguishable ligands.

The molecules are assumed to be rigid, and the groups have been determined

by geometric considerations alone, the type of skeletal geometric symmetry

envisioned in each case being indicated in its point-group notation. The

number of chemically distinct permutation isomers that each molecule has is
then immediately determined by using 2.2 proposition. The group $A_n \subset S_n$
denotes the alternating group of all the even permutations in S_n.

Table 2. Examples of molecules with four distinguishable ligands

Molecule X and its racemate group $R_X = S_X \cup \bar{S}_X$, if chiral; its chemical identity group S_X, if achiral	Geometric object with isomorphic point group symmetry	$\lvert S_X \rvert$	$\dfrac{24}{\lvert S_X \rvert}$
$L_1L_2L_3NH^\bullet$ + trace $L_1L_2L_3N$ $\qquad S_X = S_4$	none	24	1
$R_X = A_4 \cup \bar{A}_4$	Tetrahedron, T_d	12	2
$S_X=\{e,(13),(24),(12)(34),$ $(13)(24),(14)(23),$ $(1234),(1432)\}$	Square, D_{4h}	8	3
$S_X=\{e,(12),$ $(13),(23),$ $(123),(132)\}$ $(\equiv S_3)$	Regular Triangle D_{3h}	6	4
$S_X=\{e,(12)(34),$ $(13)(24),(14)(23)\}$	D_{2h} rectangle	4	6
S_X: same as above	see above	4	6

Tab. 2 cont'd.

Structure	Permutation group	Geometry / Symmetry		
allene (1,2 on left C, 3,4 on right C)	$R_X=\{e,(12)(34),(13)(24),(14)(23)\}$ $\cup \{(12),(34),(1324),(1423)\}$	D_{2d}	4	6
(fluoranthene-type structure, positions 1,2,3,4) $\longleftrightarrow$ $S_X = \{e,(13),(24),(13)(24)\}$		Rhombus D_{2h}	4	6
$C{=}N^+$ (1,2 on C; 3,4 on N) $S_X=\{e,(12)(34)\}$		Regular Trapezoid C_{2v}	2	12
stilbene-type $C{=}C$ (1,2 and 3,4) $S_X=\{e,(13)(24)\}$		C_{2h}	2	12
(tetrahedral center, 1,2,3,4) $R_X=\{e,(12)(34)\}$ $\cup \{(12),(34)\}$		C_{2v}	2	12

5. Isomerizations

The notion of a chemical identity group for a given molecule or compound is a special case of a more general concept, the chemical identity
group of a set of compounds whose molecules belong to the same family of
permutation isomers (see II,1 and IV).

To make matters precise,

5.1 Definition. Let $Q = \{A_1, \ldots, A_n\}$ be a set of isomers belonging to a
family of permutation isomers. A ligand permutation is said to preserve the
chemical identity of the system Q if, when applied to each $A_i \in Q$, it either preserves the chemical identity of A_i, or converts it to some $A_j \in Q$.

It turns out that the set of ligand permutations preserving the chemical identity of a given system Q is always a (possibly trivial) group
D[Q], called the generalized chemical identity group (or the Dieter group)
of the system Q. Observe that D[Q] reduces to the chemical identity group
if Q consists of only one member, and to the racemate group if Q is the
racemate $Q = \{A, \overline{A}\}$.

This group is used to discuss isomerizations. To illustrate the con-
cept in the general case: we are given an isomerization $A_1 \rightleftarrows A_2 \rightleftarrows \cdots \rightleftarrows A_n$
which we assume proceeds through some unknown common intermediate X, or
ensemble of intermediates X, and the problem is to determine the species X.
It is plausible to assume that, whatever X may be, any ligand permutation
that preserves the chemical identity of all reactants, or which interconverts the members of the system $\{A_1, \ldots, A_n\}$, should preserve the chemical
identity of X. Since $D[A_1, \ldots, A_n] \subset S_n$ is precisely the set of ligand per-

mutations doing this, and since it is also a group, we define $D[A_1, \ldots, A_n]$ to be the chemical identity group of the intermediate X. This characterizes the species X at the level of the chemical identity group; a geometric representation compatible with $D[A_1, \ldots, A_n]$ and the chemistry of the set $\{A_1, \ldots, A_n\}$ of stereoisomers can then be sought (see IV, 4).

5.2 Example. Given that a molecule of a tertiary amine or phosphine with a trigonal pyramidal skeleton interconverts with its enantiomer (see II, 1), we seek a mechanism for this interconversion. The ligand permutations $\{e, (123), (132)\}$ all preserve the chemical identity of A and $\bar{A}$, whereas the permutations (12), (13), and (23) interconvert A and $\bar{A}$; the intermediate species X therefore has chemical identity group $D[A, \bar{A}] = S_3$. As we have seen in II, 3, this is compatible with a mechanism in which the intermediate X has a planar skeleton with D_{3h} symmetry, meaning that the chemical identity group S_3 can represent a flexible skeleton in which the central atom oscillates above and below a coplanar arrangement of the ligands.

We can derive a general guiding principle for determining the nature of isomerization mechanisms. Say that we want to determine an isomerization mechanism for two (or more) members of a family of permutation isomers. Since the chemistry involved is known, we need only determine the Dieter group of the intermediary X. If it is not trivial, this can delineate a geometry G of an intermediary (or intermediary system) compatible with the required isomerization; if the Dieter group is trivial then, generally speaking, no non-trivial isomerization mechanism is possible. This principle in fact illuminates the Berry pseudorotation (BPR)/turnstile rotation (TR) controversy (see VII,2.1 and 2.2): the BPR [6] picks one set of two isomers for which the Dieter group is non-trivial, whereas the TR [15] picks a set of six isomers with non-trivial Dieter group. From our viewpoint, both processes are equally feasible. In practice, one or the other will prevail, depending on the properties of the individual system being considered [15,19].

Note that in the examples given we have different ways of regarding tertiary amines: in example 2.4 we obtained the chemical identity group by chemical considerations, in 4.1 we treated them as a racemic mixture and in 5.2 as an isomerization process. These groups are all the same; their interpretation alone is different, and the interpretation to be used depends on the nature of the problem.

References

[1] J. Dugundji, D. Marquarding and I. Ugi, Chemica Scripta 9, 74 (1976); 11, 17 (1977); J. Dugundji, J. Showell, R. Kopp, D. Marquarding and I. Ugi, Isr. J. Chem. 20, 20 (1980).
[2] R. Kopp, Doctoral Thesis, Technical University, München 1979.
[3] J. Gasteiger, P. D. Gillespie, D. Marquarding and I. Ugi, Topics Curr. Chem. 48, 1 (1974).
[4] J. Dugundji, R. Kopp, D. Marquarding and I. Ugi, Topics Curr. Chem. 75, 165 (1978).
[5] I. Ugi, D. Marquarding, H. Klusacek, G. Gokel, Angew. Chem. 82, 741 (1970); Angew. Chem. Int. Ed. 9, 703 (1970); in this paper the term and concept "permutation isomer" was explicitly used for the first time.
[6] R. S. Berry, J. Chem. Phys. 32, 933 (1960).
[7] The Schoenflies notation of point group symmetries is very well explained in ref.[8].
[8] R. McWeeny, "Symmetry", Pergamon, London 1962, p. 54.

[9] G. Polya, Compt. Rend. Acad. Sci. Paris 201, 1176 (1935); 202, 155
     (1936); Vierteljschr. Naturforsch. Ges. Zürich 81, 243 (1936); Z.
     Krystallogr. (A) 93, 464 (1936); Acta Math., 68, 145 (1937).

[10] N. G. De Bruijn, Koninkl. Ned. Akad. Wetenshap Proc. Ser. A62, 59
     (1959); in "Applied Combinatorial Mathematics", E. F. Beckenbach, ed.,
     p. 144, Wiley, New York 1964; Nieuw Arch. Wiskunde (3) 18, 61 (1970).

[11] E. Ruch, W. Hässelbarth, B. Richter: Theoret. Chim. Acta 19, 288
     (1970); W. Hässelbarth and E. Ruch, Theor. Chim. Acta, 29, 259 (1973);
     W. Hässelbarth, E. Ruch, D. J. Klein, T. H. Seligman "Group
     Theoretical Methods in Physics", ed.: R. T. Sharp, B. Kolman, Academic
     Press, New York 1977, p. 617.

[12] W. G. Klemperer, J. Chem. Phys. 56, 5478 (1972); J. Amer. Chem. Soc.
     94, 6940, 8360 (1972); 95, 380, 2105 (1973); Inorg. Chem. 11, 2668
     (1972).

[13] G. J. Nourse, Proc. Nat. Acad. Sci. USA 72, 2385 (1975).

[14] J. Hinze, ed., "The Permutation Group in Physics and Chemistry",
     Springer Verlag, Heidelberg 1979; see also: J. Brocas, M. Gielen and
     R. Willem, "The Permutational Approach to Dynamic Stereochemistry",
     McGraw-Hill, New York 1983.

[15] see e. g.: P. Gillespie, P. Hoffmann, H. Klusacek, D. Marquarding,
     S. Pfohl, F. Ramirez, E. A. Tsolis and I. Ugi, Angew. Chem. 83, 691
     (1971); Angew. Chem. Int. Ed. 10, 687 (1971); A. T. Balaban, ed.:
     "Chemical Applications of Graph Theory", Academic Press, London 1976;
     J. G. Nourse, J. Amer. Chem. Soc., 99, 2063 (1977).

[16] R. S. Cahn, C. K. Ingold and V. Prelog, Angew. Chem. 78, 413 (1966);
     Angew. Chem. Int. Ed. 5, 385 (1966); V. Prelog and G. Helmchen, Angew.
     Chem. 94, 614 (1982); Angew. Chem. Int. Ed. 21, 567 (1982).

[17] W. Schubert and I. Ugi, J. Amer. Chem. Soc. 100, 37 (1978); Chimia 33,
     183 (1979).

[18] see also: J. Blair, J. Gasteiger, C. Gillespie, P. D. Gillespie and
     I. Ugi, Tetrahedron 30, 1845 (1974); W. T. Wipke and T. M. Dyott, J.
     Amer. Chem. Soc. 96, 4825 (1974).

[19] F. Ramirez and I. Ugi, in: "Advances in Physical Organic Chemistry",
     ed.: V. Gold, Academic Press, London 1971, p. 25; J. Dugundji, P. D.
     Gillespie, D. Marquarding, I. Ugi and F. Ramirez, in: "Chemical
     Applications of Graph Theory", ed.: A. T. Balaban, Academic Press,
     London 1976, p. 107, and references therein.

[20] By "chemically distinguishable" is meant: distinguishable by any observation method used in chemistry or physicochemistry. Thus isotopically different atoms would be chemically distinguishable.

THE ASYMMETRIC CARBON ATOM REVISITED

In this chapter, we illustrate our general approach to stereochemical problems, and the use of the concepts developed in the previous chapter, by studying the asymmetric carbon atom [1]: we first determine its chemical identity group, and then find that the group is compatible with the usual tetrahedral valence skeleton. It will be pointed out that even in the case of this molecule, which can be represented by a simple geometric model, a purely geometric view of its stereochemical features has inherent logical difficulties, and that these difficulties disappear when the chemical identity viewpoint is adopted.

1. *Chemical Identity Group of the Asymmetric Carbon Atom*

We shall base our discussion on the chemical evidence available to Le Bel [2] and van't Hoff [3], that all the ways of attaching four chemically distinguishable ligands to a carbon atom give molecules of exactly two distinct enantiomeric compounds.

We now express the known chemical facts in terms of ligand permutations on a model E of C and determine the chemical identity group of an "asymmetric carbon" C. Since there are four ligands, the chemical identity group S_C is a subgroup of S_4. As observed in IV.2, the number of stereoisomers of C that can be formed by all possible ways of attaching the ligands to E is in a 1-1 correspondence with the family of left cosets of S_C in S_4; and since it is known that there are only two stereoisomers possi-

43

ble, this family of cosets consists of S_C and a single coset $\bar{S}_C$ of S_C. (We remind the reader that S_C and $\bar{S}_C$ have the same number of elements, and that $S_C \cup \bar{S}_C = S_4$.) Because S_4 has order 24, it follows that $S_C \subset S_4$ must be a subgroup of order 12. The only subgroup of order 12 in S_4 being A_4, the alternating group (i. e. all the even permutations in S_4), we conclude that $S_C = A_4$ and that its coset $\bar{S}_C$ consists of all the odd permutations belonging to S_4. In particular, every even permutation of the ligands of E represents the chemical compound C (see II.2), and every odd permutation represents the enantiomer $\bar{C}$ (see II,2.2).

Observe also that, because C is known to be chiral and because its enantiomer $\bar{C}$ belongs to the same family of permutation isomers, the skeleton is achiral, i. e. the chirality of C is due to different placements of the ligand on the same skeleton. The compound C has a racemate group (see II.4) which contains S_C as a subgroup of index 2; the only such group containing A_4 being S_4 itself, we conclude that the enantiomer coset $\bar{S}_C$ consists of all the odd permutations in S_4, and therefore (what we already know) that each such permutation converts C to its enantiomer $\bar{C}$.

2. *Geometrical Interpretation of the Asymmetric Carbon Atom*

With this knowledge of the chemistry and the chemical identity group S_C of the asymmetric carbon atom C and its enantiomer $\bar{C}$, we are ready to consider some of the conceivable stereochemical interpretations (i. e. some of the appropiate G in the equation $S_X = F(C,G)$ of II,3).

Assume first that the skeleton is rigid. The group A_4 has a classical and well known interpretation as the T_d point group symmetry of a tetrahedron [4]. With this skeletal symmetry, the central carbon atom is located at the center of a tetrahedron whose vertices are occupied by the ligands 1,...,4. Then any even permutation (i. e. member of S_C) of the idealized ligands of a model E leads to a rotated form of E (i. e. can be brought to coincidence with E by rotation of the entire molecule). For example, the even permutation (123) represents a 120° rotation of E = 1a about an axis passing through the central atom and ligand 4; thus 1a gives a molecule 1b:

1a   1b $\overline{1a}$

The odd ligand permutations (i. e. members of the enantiomer coset $\overline{S_C}$) convert 1 into $\overline{1}$, the enantiomer of 1.

Thus, the chemical identity group A_4 for the asymmetric carbon atom is compatible with the customary geometrical representation of that molecule: both serve to explain the observed chemical behaviour.

However, instead of assuming that the skeleton is rigid, we could assume that the tetracoordinate skeleton has a D_{2d} allene type symmetry. Such a skeleton would qualify, if the molecular system C consisted of an equimolar mixture of three rapidly equilibrating distinct molecules 2a - 2c, and $\overline{C}$ consisted of $\overline{2a}$ - $\overline{2c}$:

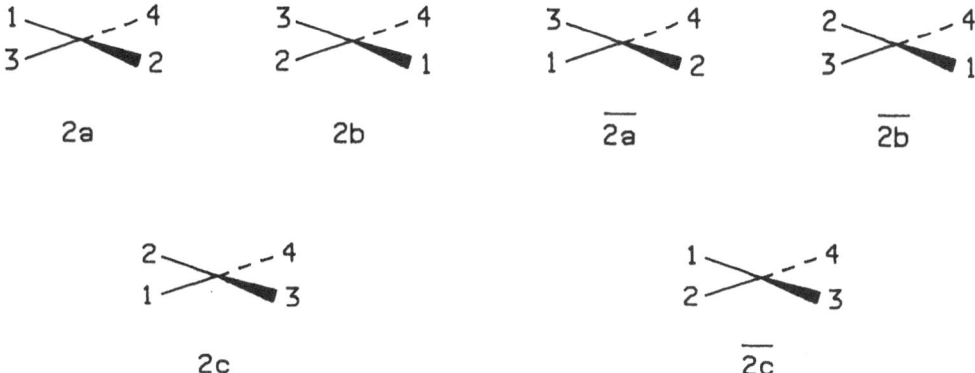

The chemical identity of the ensemble C so well as that of $\overline{C}$ is maintained by the ligand permutations belonging to S_C, regardless of whether or not the individual members of C (and of $\overline{C}$) interconvert. The ligand permutations in the enantiomer coset (i. e. the odd permutations) lead from C to $\overline{C}$ and vice versa.

As our discussion shows, the rigid asymmetric carbon atom is not the only molecular model compatible with the experimental evidence that was available to van't Hoff and Le Bel. However, even if the asymmetric carbon atom had been endowed with a nonrigid skeleton (like many presently known species) the fact that it behaves as though it had a rigid tetrahedral skeleton would lead to a model permitting valid interpretations of compounds with asymmetric carbon atoms, despite the "wrong geometry". In fact, a specific case in which a geometrically wrong model is often used for arriving at correct conclusions is the time average model of cyclohexane [5].

3. *Contrast of the Geometrical and Permutational Interpretations*

The simple and extremely attractive idea that the observed chemical behaviour of an asymmetric carbon atom can be explained by a T_d point-group symmetry of the molecular skeleton has a serious logical shortcoming: such a symmetry is rarely found on real molecules. In fact, in an asymmetric carbon atom the bond angles generally deviate from the idealized 109° $33'$, due to differences in the interaction of the pairwise different ligands; in addition, the bond lengths between the central atom and the ligands are also not all the same. For example, for bond angles in CHBrClF it is known that ∢ HCF < HCCl < HCBr < ClCBr and for bond lengths, that C–H < C–F < C–Cl < C–Br.

Thus, although the asymmetric carbon atoms behave as if they had an idealized T_d skeleton, they in fact never have that idealized T_d skeletal symmetry. Since there is no such thing as an "approximate symmetry" (how "approximate"? Is it more "approximate" to a symmetry not yet mentioned? [6]) the geometrically based approach has a difficulty in justifying the use of symmetry considerations which work flawlessly in practice, but appear to have no justification based on known facts about the variations in skeletal geometry.

This logical inconsistency is entirely avoided by using our permutational approach to stereochemistry. As we have seen in III.1, the geometry of the asymmetric carbon atom is of little use in determining its chemical identity group: the basic consideration was the variation in the chemical identity of the molecule under permutations of the ligands; and it was information from chemistry, rather than from geometry, that enabled us to determine the group S_C. By regarding the deformations in bond length and bond angle at the valence skeleton of the central atom to "move along" with the ligands undergoing permutation, the skeleton and its symmetries can be totally neglected. The asymmetric carbon atom can therefore be interpreted in terms of ligand permutations without any idealization or approximation standpoint. From this viewpoint, the classical interpretation of the asymmetric carbon atom is so successful, although the underlying geometrical ideas (see III,2) are definitely not valid, because the chemical identity group S_C and the rotational symmetry point-group of the idealized tetrahedral skeleton are isomorphic [4].

The above discussion can be summarized in the following way: once S_C has been determined, many abstract models can in general be constructed having the group S_C as a group of symmetries; any of these abstract models then indicates, by geometric considerations, what a given ligand permutation does to the chemical identity of the given molecule. In these terms, the usual picture of the asymmetric carbon atom does not depict any reality about the geometrical structure of that atom; it is simply a mnemonic device to indicate whether or not a given ligand permutation will change the chemical identity of the given compound; and this is probably the basic reason that the flawed model for the asymmetric carbon atom works so flawlessly in practice.

References

[1] J. Weyer, Angew. Chem. 86, 604 (1974); Angew. Chem. Int. Ed. 13, 591
     (1974); see also: K. Mislow and J. Siegel, J. Amer. Chem. Soc.
     (in press).

[2] J. A. Le Bel, Bull Soc. Chim. France [2] 22, 337 (1974).

[3] J. H. van't Hoff, "Voorstel tot utbreiding der tegenwoordig in de scheikunde gebruikte structurenformules in de ruimte", Greven, Utrecht 1874.

[4] R. Mc Weeny, "Symmetry", Pergamon, London 1962, p. 54.

[5] J. E. Leonard, G. S. Hammond and H. E. Simmons, J. Amer. Chem. Soc. 97, 5052 (1975).

[6] At our institute P. Lemmen and R. Baumgärtner have implemented a computer program for answering just these questions. Some rather interesting results will be published soon.

PART II

THE MATHEMATICAL THEORY OF THE CHEMICAL IDENTITY GROUP

FAMILIES OF PERMUTATION ISOMERS

1. *Permutation Isomers*

The molecules of a pure and uniform chemical compound X are all chemically identical; although they may differ in shape at a given time, they
interconvert spontaneously under the observation conditions. In chemistry
it is customary to represent the stereochemistry of X by drawing some pro-
jection formula of a representative "molecular situation" belonging to X.

We review the notion of a family of permutation isomers [1] of X. A
model for X is chosen and conceptually dissected into a set of ligands L
and a skeleton, in a fashion that is appropriate for the problem [2]. The
ligands are those atoms, or polyatomic residues that can be interchanged;
what remains after the removal of the ligands is called the skeleton.

This model is called a reference model E for X. Any desired redistribution of the ligands on the skeletal sites can then be specified simply by
a permutation of the ligands on the reference model E [3,4]. Each such
rearrangement converts E to a model representing some permutation isomer;
the set of all permutation isomers obtainable is called the family $J_X(L)$ of
permutation isomers of X relative to L [1].

Note that in order to define a family of permutation isomers of X, the
conceptual dissection of X into a skeleton and a set of ligands must be
specified; the family $J_X(L)$ depends on the dissection that is chosen. For
example, in the compound X represented by **1**

1

a dissection into a methane skeleton with $L = \{Cl, COOH, F, CHF_2\}$ would be appropriate in some studies emphasizing configuration, whereas its representation as an ethane skeleton with $L' = \{Cl, F, COOH, F, F, H\}$ would be appropriate in the study of its conformation [2]; the families $J_X(L)$ and $J_X(L')$ are not the same. We remind the reader that permutation isomers need not be stereoisomers, and vice versa.

To discuss families of permutation isomers, we need some notation. It is no loss of generality to label the ligands $1, 2, \ldots, n$ so that they are all mathematically (but not necessarily chemically) distinguishable; and to give our considerations the broadest scope, we assume that all the different ways of placing the ligands on the skeletal sites give chemically meaningful molecules. Now let E be a reference model for X, and let $\lambda = (n_1, \ldots, n_k)$ be any permutation in SymL, the symmetric group on $|L|$ letters. Then λE will denote the model obtained from E by the interchange $n_1 \to n_2 \to \ldots \to n_k \to n_1$ of the ligands, i. e. n_1 replaces n_2 etc.

Performing the permutation μ on the ligands of λE is denoted by $\mu(\lambda E)$; the same molecule is obtained by applying the product $\mu \cdot \lambda$ (in the order written!) of the permutations to the ligands of the reference model E; i. e. $\mu(\lambda E) = \mu \cdot \lambda E$. Thus, $(23)[(142)E] = (1423)E$.

$$E = \quad 3\overset{1}{\underset{2}{\diagdown}}{}^{\!-\!4} \xrightarrow{\;(142)\;} \quad 3\overset{2}{\underset{4}{\diagdown}}{}^{\!-\!1} \xrightarrow{\;(23)\;} \quad 2\overset{3}{\underset{4}{\diagdown}}{}^{\!-\!1}$$

$$E \qquad\qquad (142)E \qquad\qquad (23)[(142)E]=(1423)E$$

Note that by denoting successive operations on the ligands in "functional" fashion, their combined effect amounts to the usual product of the permutations (in the order that they are written!) performed directly on the reference model.

Using this notation, we summarize the discussion in

1.1 <u>Definition</u>. Let E be a reference model for compound X, and let L be the set of ligands which expresses the mode of dissection. By the family $J_X(L)$ of permutation isomers of X relative to L is meant the set of all isomers represented by the models in $\{\mu E | \mu \in \text{SymL}\}$. The isomer X represented by the model E is called the reference isomer of this family.

It is frequently more convenient to deal directly with the molecule μE rather than with the isomers that they represent. We call the set of models $\{\mu E | \mu \in \text{SymL}\}$ a family of <u>permuted models</u> with reference model E, and denote this set by $P_E(L)$; thus, each model in $P_E(L)$ represents some isomer in $J_X(L)$. Observe that if all the ligands are chemically distinguishable, then $P_E(L)$ has exactly $|L|!$ distinct models, whereas the set $J_X(L)$ may have far fewer distinct members.

There is a significant advantage in using only ligand permutations on a fixed model E to represent the members of $J_X(L)$: since emphasis is entirely on ligand permutations, no idealized skeletal geometry has to be assumed. In fact, since the process involves only a set of sites (which need not always remain in the same place) and a set of ligands to be attached to those sites, the notion of permutational isomer is meaningful even if the skeleton is flexible, or not contiguous, or even when the skeleton is taken to be an ensemble of (possibly different) skeletons. Indeed, permutational concepts apply equally well, in a formal way, to individual molecules and to ensembles of molecules.

2. The Fundamental Theorem on Molecules with all Ligands Chemically

Distinguishable

Let X be a chemical compound with reference model E. We assume that the ligands L are all chemically distinguishable, and seek to determine the number of distinct permutation isomers in $J_X(L)$.

For each $\mu \in \text{SymL}$, the model μE represents some member of $J_X(L)$; but distinct $\lambda E, \mu E$ may well represent the same chemical compound. For example, one may be simply a rotated form of the other:

E   (13)(24)E

We now introduce into SymL a relation "~" determined by the chemical features of E with

2.1 Definition. $\mu \sim \lambda$ if μE is chemically identical to λE

It is obvious on semantic grounds that ~ is in fact an equivalence relation in SymL, i. e. that it is reflexive, symmetric, and transitive; it therefore decomposes SymL into mutually exclusive equivalence classes, two permutations λ,μ belonging to the same class if and only if λE is chemically identical to μE. Each equivalence class in SymL therefore represents a distinct isomer, and the number of equivalence classes is the number of distinct isomers in $J_\chi(L)$. With this, our problem reduces to determining the equivalence classes in SymL.

It is a generally accepted chemical fact that, if all the ligands of E are chemically distinct, then whenever $\lambda E,\mu E$ are chemically identical, so also will be the molecules $\alpha\lambda E,\alpha\mu E$ for each $\lambda \in$ SymL. (For example, if $\lambda E,\mu E$ are "rotated" forms of one another, then the same ligand permutation applied to both of them will give molecules that are "rotated" forms of one another.) This chemical fact indicates that the equivalence relation ~ in SymL is related to the group operation in SymL by the

2.2 **Stereochemical Axiom.** If all the ligands of E are chemically distinguishable, then whenever $\lambda \sim \mu$ also $\alpha\lambda \sim \alpha\mu$ for each $\alpha \in$ SymL.

This axiom is the starting point of our theory, and can be regarded as expressing the chemistry/geometry interplay in stereochemistry; indeed, all the results we get also apply formally to any molecule (or ensemble of molecules) taken with a "ligand/skeleton" decomposition.

The stereochemical axiom leads immediately to the following fundamental theorem of our theory, which shows among other things that the use of group theory to describe stereochemical phenomena and properties is inherent in the nature of stereochemistry:

2.3 Theorem.   Let $P_E(L)$ be a family of permuted models, with all ligands chemically distinguishable. Let "~" be the equivalence relation 2.1 in SymL, and for each $\lambda \in$ SymL, let [λ] denote its equivalence class. Then

1. The equivalence class [e] containing the identity permutation is a group.

2. The equivalence class [λ] is the left coset λ[e] in SymL.

3. The number of equivalence classes is the index [SymL:[e]] of [e] in SymL.

Proof

Ad 1).   Let $\lambda,\mu \in$ [e] so that $\lambda \sim$ e and $\mu \sim$ e; we are to show that $\lambda^{-1}\mu \in$ [e]. By symmetry and transitivity, we find $\mu \sim \lambda$ and, from 2.2, that $\lambda^{-1}\mu \sim \lambda^{-1}\lambda$ = e. Thus, $\lambda^{-1}\mu \in$ [e], and [e] is therefore a group.

Ad 2).   We first show that [λ] $\subset \lambda$[e]: if $\mu \in$ [λ], then $\mu \sim \lambda$ so $\lambda^{-1}\mu \sim$ e, therefore $\lambda^{-1}\mu \in$ [e], and so $\mu \in \lambda$[e]. By reversing the steps of this argument, we find that if $\mu \in \lambda$[e], then $\mu \in$ [λ], and therefore that λ[e] $\subset$ [λ]. We conclude that [λ] = λ[e].

Ad 3).   This is an immediate consequence of (2).

    The group [e] that we have found is basic in our development of the subject. Note that the equivalence relation ~ is determined by the observed behaviour of X; with varying observation conditions and/or varying data about X, the group [e] will in general also vary. Observe also that the group [e] is defined only when all the ligands are chemically distinguish-able.

To make the 2.3 Theorem useful, it is necessary to describe the elements of [e] in more familiar terms. For this purpose, observe that by the definition of the equivalence relation ~, we have $\lambda \in [e]$ if and only if λE is chemically equivalent to E, i. e. if and only if the permutation λ on the ligands of E preserves the chemical identity of X. Therefore

2.4 **Corollary**. The group [e] is precisely the set of all permutations $\lambda \in SymL$ that preserve the chemical identity of X. We call [e] the chemical identity group of X, and denote it by S_X.

It is this characterization of [e], as the set of all ligand permutations preserving the chemical identity of X, that provides a means for its determination in any given experiment; as described in II, this can be accomplished by using, say, NMR studies, or observed number of isomers, or by an assumed skeletal geometry/chemistry interplay.

Using this terminology, we can rephrase 2.3 Theorem in a manner more convenient for our later applications (see II,2.2).

2.5 **Theorem**. Let X be a chemical compound with reference model E having all ligands chemically distinguishable. The set of all permutations in SymL that preserve the chemical identity of X form a group S_X, called the chemical identity group of X. The number of distinct permutation isomers of X is $[SymL:S_X]$; and two permutations λ, μ produce chemically identical $\lambda E, \mu E$ if and only if λ, μ belong to a common left coset αS_X.

Because of this theorem, we call the left cosets λS_X the permutation isomers of X, which amounts to using the left cosets as a nomenclature (see VIII,6) for the isomers of X. If $\mu \in \alpha S_X$, we say that the model μE belongs to (or represents) the isomer αS_X. In this terminology, the reference model E belongs to the reference isomer X which is represented by S_X.

3. The Chemical Identity Group of an Isomer

Let $J_X(L)$ be a family of permutation isomers with all ligands chemically distinguishable, and with reference model E belonging to the reference isomer X having the chemical identity group S_X. We shall extend the idea of 2.4 to define the chemical identity group of each isomer.

Each $\alpha \in$ SymL operates on the set of cosets $\{\lambda S_X\}$ by the rule $\alpha(\lambda S_X) = \alpha\lambda \cdot S_X$. This operation of α is well-defined, i. e. the value depends on the coset λS_X itself, rather than on the way it is written, since $\lambda S_X = \mu S_X$ gives $\alpha\lambda S_X = \alpha\mu S_X$. Observe that this operation summarizes the effect of the ligand permutation α on all the models γE belonging to the isomer λS_X: given any such model γE, the model $\alpha\gamma E$ belongs to the isomer $\alpha\lambda S_X$, because $\gamma \sim \lambda$ gives $\alpha\gamma \sim \alpha\lambda$. With this, we explicitly formulate the basic

3.1 Definition. A permutation $\alpha \in$ SymL is said to preserve the chemical identity of the isomer λS_X if $\alpha\lambda S_X = \lambda S_X$

and show

3.2 Theorem. Let $J_X(L)$ be a family of permutation isomers, with all ligands chemically distinguishable, with reference model E, and reference isomer X having chemical identity group S_X. The set of all permutations preserving the chemical identity of the isomer λS_X is precisely the group $\lambda S_X \lambda^{-1} \subset SymL$.

Proof We have $\alpha \lambda S_X = \lambda S_X$ if and only if $\lambda^{-1}\alpha\lambda \in S_X$ or, equivalently, if and only if $\alpha \in \lambda S_X \lambda^{-1}$. This completes the proof.

By analogy with 2.4, we call $\lambda S_X \lambda^{-1}$ the chemical identity group of the isomer λS_X. On the molecular level, 3.2 means that for each model μE belonging to the isomer λS_X, the permutations $\lambda S_X \lambda^{-1}$ applied to μE give models chemically identical to μE, so we also call the group $\lambda S_X \lambda^{-1}$ the chemical identity group of the isomer represented by λS_X. Note once again that the concept of a chemical identity group is defined only when all the ligands are chemically distinguishable.

We now investigate the ability of the chemical identity group to distinguish between distinct permutation isomers of a compound X. From 3.2, the set of chemical identity groups of the permutation isomers of X is a single conjugacy class of subgroups of SymL, with the reference isomer X carrying the group S_X, and the permutation isomer λS_X the group $\lambda S_X \lambda^{-1}$. The number of chemically distinct permutation isomers of X is, according to 2.5, exactly $[SymL:S_X]$; the number of distinct conjugates to S_X is $[SymL:N(S_X)]$, where $N(S_X)$ is the normalizer of S_X in SymL (see Appendix). Therefore

3.3 Theorem Let $J_X(L)$ be a family of permutation isomers, with S_X the chemical identity group of X. The chemical identity groups of the permutation isomers form a single conjugacy class of groups in SymL. For each conjugate $\lambda S_X \lambda^{-1}$, there are exactly $[N(S_X):S_X]$ distinct permutation isomers having that group as chemical identity group.

<u>Proof</u>   Because $S_X \subset N(S_X)$, the result is immediate from the equation $[SymL:S_X] = [SymL:N(S_X)] \cdot [N(S_X):S_X]$.

Thus, for example, there will be exactly $[N(S_X):S_X]$ distinct permutation isomers having the same chemical identity group, S_X, as the reference isomer X. In particular, whenever $[N(S_X):S_X] \geq 2$ and mathematical results involving chemical identity groups are interpreted in chemical terms, some attention must be given to the chemical facts involved in order to identify the exact isomer being described (see VII,1, 2.2 and 2.3).

4.   *The Chemical Identity Group of a Set of Permutation Isomers*

The notion of a chemical identity group of a single isomer is a special case of a more general concept, that of the chemical identity group of a set of isomers belonging to the same family of permutation isomers.

<u>4.1 Definition</u>. Let $Q = \{\lambda_1 S_X, \ldots, \lambda_n S_X\}$ be a set of distinct permutation isomers in $J_X(L)$, where all the ligands are chemically distinguishable. A permutation $\alpha \in SymL$ is said to preserve the chemical identity of the system Q if for each $\lambda_i S_X \in Q$ the $\alpha \lambda_i S_X$ is also a member of Q.

Thus, the permutations α that preserve the chemical identity of the given system Q lead to isomers that also belong to Q. This set of per-

mutations always forms a (perhaps trivial) group D[Q]: If σ,τ are two

permutations in D[Q], then for each $\lambda_i S_X$, we have $\sigma\lambda_i S_X$ is some $\lambda_j S_X$,

and $\pi\lambda_j S_X$ is some $\lambda_k S_X$, so $\pi\sigma\lambda_i S_X = \lambda_k S_X$; thus, the composition of

two $\sigma,\pi \in$ D[Q] also belongs to D[Q], and since SymL is a finite group,

we conclude (Appendix, 1.2) that D[Q] is a group.

4.2 **Definition**. The group D[Q] $\subset$ SymL of all permutations preserving the

chemical identity of the system Q is called the chemical identity group

(or: Dieter group) of the system Q.

The importance of this concept stems from the fact pointed out before,

that D[Q] can serve as the chemical identity group of an intermediate

species (or set of species) in a hypothesized isomerization mechanism.

We now obtain explicit formulas for D[Q], and derive some conditions

under which D[Q] will not be trivial.

4.3 **Theorem**. Let $J_X(L)$ be a family of permutation isomers where the re-

ference isomer X has chemical identity group S_X, and let $Q=\{\lambda_1 S_X,\ldots,\lambda_n S_X\}$

be a given set of distinct isomers. Let

$$T = \lambda_1 S_X \cup \ldots\ldots \cup \lambda_n S_X$$

be the union of the cosets belonging to Q. Then

$$D[Q] = T\lambda_1^{-1} \cap T\lambda_2^{-1} \cap \ldots \cap T\lambda_n^{-1}$$

independently of the representatives $\lambda_1,\ldots,\lambda_n$ that are used.

Proof To say that μ preserves the chemical identity of the system Q means that for each $i = 1, \ldots, n$, we have $\mu\lambda_i$ belonging to some coset $\lambda_j S_X$

or, in other words, that $\mu\lambda_i \in \lambda_1 S_X \cup \ldots \cup \lambda_n S_X = T$ for each

$i = 1, \ldots, n$. Conversely, if μ is any permutation with $\mu\lambda_i \in T$ for each

$i = 1, \ldots, n$, then μ preserves the chemical identity of the system Q.

Thus, $\mu \in D[Q]$ if and only if $\mu\lambda_i \in T$ for each $i = 1, \ldots, n$, i. e. if

and only if $\mu \in T\lambda_i^{-1}$ for each $i = 1, \ldots, n$. Therefore $D[Q] = \bigcap_1^n T\lambda_i^{-1}$

as asserted. This depends on the cosets themselves, rather than on the

representatives λ_i we have used: for if γ_i is any other member of $\lambda_i S_X$,

then $\lambda_i^{-1} \gamma_i \in S_X$ so that $T\lambda_i^{-1} \gamma_i = T$ and therefore $T\lambda_i^{-1} = T\gamma_i^{-1}$.

This completes the proof.

The group $D[Q]$ can be represented in another way, which emphasizes the

effect each $\mu \in D[Q]$ has. Starting with the $n \times n$ array

$$
D[Q] = \begin{vmatrix}
\lambda_1 S_X \lambda_1^{-1} & \lambda_2 S_X \lambda_1^{-1} & \cdots & \lambda_n S_X \lambda_1^{-1} \\
\lambda_1 S_X \lambda_2^{-1} & \lambda_2 S_X \lambda_2^{-1} & \cdots & \lambda_n S_X \lambda_2^{-1} \\
\cdots & \cdots & & \cdots \\
\lambda_1 S_X \lambda_n^{-1} & \lambda_2 S_X \lambda_n^{-1} & \cdots & \lambda_n S_X \lambda_n^{-1}
\end{vmatrix}
$$

we note that $D[Q]$ is the intersection of the unions of the rows. By the

distributive law, this will be the union of the intersections formed by

taking one term from each row. However, any such intersection that involves

two terms in the same column will be empty, because all the terms in each

column are fixed translations of distinct right cosets of S_X: if

$\lambda_i^{-1} \in S_X\lambda_j^{-1}$, then $\lambda_i^{-1}\lambda_j \in S_X$ so $\lambda_j \in \lambda_i S_X$ which is ex-

cluded by our hypothesis that the $\lambda_i S_X$ are distinct isomers. Thus, $D[Q]$

will be the union of all the intersections formed by taking n terms of the array, no two of which are in the same row or column, i. e. D[Q] is the permanent of the above array (where product $=$ ∩ and sum $=$ ∪). Each term of this permanent has a chemical interpretation: for example, the elements (if any) in the term

$$\lambda_2 \, S_X \, \lambda_1^{-1} \; \cap \; \lambda_1 \, S_X \, \lambda_2^{-1} \; \cap \; \lambda_3 \, S_X \, \lambda_3^{-1} \; \cap \; \ldots \; \cap \; \lambda_n \, S_X \, \lambda_n^{-1}$$

are the members μ of D[Q] which satisfy $\mu\lambda_1 S_X = \lambda_2 S_X$, $\mu\lambda_2 S_X = \lambda_1 S_X$ and $\mu\lambda_i S_X = \lambda_i S_X$ for all the remaining i. Although this representation of D[Q] as a permanent is easy to work with when $|Q| \leq 3$, it is unwieldy for larger $|Q|$.

There is still another description of D[Q] that works directly with a system of representatives of the cosets involved.

4.4 Proposition   Let $Q = \{\lambda_1 \, S_X, \ldots, \lambda_n \, S_X\}$ be a set of distinct isomers in $J_X(L)$. Then $D[Q] = \{\mu \in \text{SymL} \mid \{\mu\lambda_i\}$ and $\{\lambda_i\}$ represent the same set Q of cosets}.

Proof   Observe first that for any $\mu \in \text{SymL}$, the $\mu\lambda_i$ and $\mu\lambda_j (i \neq j)$ cannot belong to the same coset of S_X: for if $(\mu\lambda_i)^{-1}(\mu\lambda_j) \in S_X$ this would imply that $\lambda_i^{-1}\lambda_j \in S_X$, i. e. $\lambda_j \in \lambda_i S_X$, which is excluded.

Now let $\mu \in D[Q]$ and fix any index i. Since $\mu \in T \, \lambda_i^{-1}$, we find $\mu \in \lambda_k \, S_X \, \lambda_i^{-1}$ for some k, therefore $\mu\lambda_i \in \lambda_k S_X$ for some k. This is true for each $i = 1, \ldots, n$ so, by our observation above, $\{\mu\lambda_i\}$ represents the set Q of cosets.

Conversely, if $\{\mu\lambda_i\}$ and $\{\lambda_i\}$ represent the same family of cosets Q, then for each i, the $\mu\lambda_i \in \lambda_k S_\chi$ for some k, so $\mu \in \lambda_k S_\chi \lambda_i^{-1}$ $\subset T\lambda_i^{-1}$ for each i, therefore $\mu \in D[Q]$.

Remarks

1. The Dieter group $D[Q]$ of a given Q may not be transitive on Q, i. e. there may be no $\alpha \in D[Q]$ that converts a given isomer $\lambda S_\chi \in Q$ to another given isomer $\mu S_\chi \in Q$.

2. The Dieter group $D[Q]$ may not act primitively on Q, i. e. it may be possible to decompose Q into blocks $Q_1 \cup \ldots \cup Q_s$ where $1 < |Q_i| < |Q|$ for each i, such that each $\alpha \in D[Q]$ maps each block Q_i onto some block Q_j. The blocks Q_i are called imprimitivity domains.

We now seek some conditions that will assure $D[Q] \neq \{e\}$. The 4.4 Proposition indicates a convenient method: one needs only produce some $\mu \neq e$ that permutes the $\{\lambda_i\}$, up to cosets. The most important case occurs when the coset representatives can be chosen to form a group (see VII,2.2).

4.5 Theorem   Let $Q = \{S_\chi, \lambda_1 S_\chi, \ldots, \lambda_n S_\chi\}$ be a set of isomers in $J_\chi(L)$. Assume that $\{e, \lambda_1, \ldots, \lambda_n\}$ forms a group G. Then $G \subset D[Q]$ and, in particular, $D[Q] \neq \{e\}$.

Proof   This is immediate from 4.4 Proposition, because the product of $\{e, \lambda_1, \ldots, \lambda_n\}$ with any λ_i simply permutes the system $\{e, \lambda_1, \ldots, \lambda_n\}$.

The actual calculation of $D[Q]$ can also be considerably simplified whenever the system $\{e, \lambda_1, \ldots, \lambda_n\}$ of representatives is a group. For, from

$T = S_X \cup \lambda_1 S_X \cup \ldots \cup \lambda_n S_X$ we find $\lambda_i T = T$ for each i, so

$\lambda_i T \lambda_i^{-1} = T \lambda_i^{-1}$, therefore

$$D[Q] = \bigcap_1^n \lambda_i T \lambda_i^{-1}$$

and the computation of the conjugates $\lambda_i T \lambda_i^{-1}$ is easier to perform than

that of the products $T \lambda_i^{-1}$. Moreover, sometimes the determination of D[Q]

itself is immediate:

4.6 Proposition Let $Q = \{S_X, \lambda_1 S_X, \ldots, \lambda_n S_X\}$ be a system of isomers in

$J_X(L)$ and assume that $\{e, \lambda_1, \ldots, \lambda_n\}$ is a group G. Then $D[Q] = G \cdot S_X$ if and

only if $G \cdot S_X$ is a group.

Proof If $D[Q] = G \cdot S_X$, then $G \cdot S_X$ is a group. It remains to prove the

converse. For this, we first note that $G \subset D$ by 4.5. We next show that

$S_X \subset D$: for, given any $s \in S_X$, consider the system $\{s\lambda_i\}$ of repre-

sentatives. Clearly, no two of these belong to the same coset of S_X; and

because $G \cdot S_X$ is a group, we have $S_X \cdot G = G \cdot S_X$, therefore each $s\lambda_i$ is

some $\lambda_j \hat{s}$, so each coset $s\lambda_i \cdot S_X$ is some $\lambda_j \cdot S_X$. Thus, by 4.4, we find

$S_X \subset D[Q]$. Finally, note that because $T = S_X \cup \lambda_1 S_X \cup \ldots \cup \lambda_n S_X$, we have

$T \subset G \cdot S_X = S_X \cdot G$, therefore $T \lambda_i^{-1} \subset S_X \cdot G$ for each i, consequently

$D \subset G \cdot S_X$. We have therefore shown that $G \subset D \subset G \cdot S_X$ so, by Dedekind's

rule (Appendix, 2.7), we find that

$$D[Q] = G \cdot (S_X \cap D) = G \cdot S_X$$

and this completes the proof.

5. *Involution Families and Racemate Groups*

Recall that a chiral molecule X is said to have an achiral skeleton if its enantiomer $\overline{X}$ belongs to the same family $J_X(L)$ of permutation isomers (see II,4).

Assuming that the skeleton is achiral and that all the ligands are chemically distinguishable, we seek to determine the essential features of a permutation that converts X to its enantiomer. Such a permutation ρ (called an enantiomerization) corresponds to a permutation of the ligands of X when X is reflected in a mirror, so it should satisfy $\rho^2 = e$; and since the enantiomer $\overline{X}$ is not chemically equivalent to X, we must have $\rho \notin S_X$. Moreover, the enantiomers of chemically identical species should be chemically identical so, thinking of the reflection in a mirror, any permutation of the ligands of X that preserves its chemical identity, should also preserve the chemical identity of $\overline{X}$. Therefore by 3.2, we should have $S_X = \rho \, S_X \, \rho^{-1}$, i. e. ρ should be in the normalizer $N(S_X)$ of S_X.

Guided by these considerations, we make the

5.1 Definition. Let $J_X(L)$ be a family of permutation isomers, and let S_X be the chemical identity group of the reference isomer. We say X has an involution if there is some $\rho \in \text{SymL}$ with $\rho \in N(S_X) - S_X$ and $\rho^2 \in S_X$. The isomer ρS_X is called the ρ-involution isomer of S_X.

From what we have said above, every enantiomerization is an involution. However, enantiomerizations are determined by special geometrical/chemical considerations, whereas involutions arise from purely algebraic considerations and, in general, there exist involutions that are not enantiomerizations. In fact, an achiral molecule may have an involution, and a chiral molecule can have an involution that is not an enantiomerization, as the following examples show:

Ex. 1

with $S = \{e,(13)(24),(13),(24)\}$ is achiral; but it has the involution $\rho = (1234)$

Ex. 2

with $S = \{e,(12)(34),(13)(24),(14)(23)\}$, has the enantiomerization $\rho = (12)$; it also has the involution $\rho = (13)$ which is not an enantiomerization.

We say that a molecule is chiral if it has an enantiomerization. It is easy to see that a molecule X has an involution if and only if $|N(S_X)|/|S_X|$ is even, so that this condition is necessary (but in general not sufficient) for X to be chiral.

Since the concept of an involution contains that of enantiomerization, it seems worthwhile to study chirality from this more general viewpoint. Each involution will be seen to decompose the family of permutation isomers into pairs which can be considered to be isomerizing through well-determined intermediates, and these intermediates themselves form a family of permutation isomers. In the case that the involution is an enantiomerization, the intermediates have chemical identity groups that can be regarded to be those of racemic mixtures of enantiomer pairs.

An involution ρ has the following properties, which we shall use in the sequel.

5.2 Proposition Let $\rho \in \text{Sym}L$ be an involution of X. Then $\rho \, S_X \, \rho^{-1} =$ $\rho^{-1} S_X \, \rho \; = \; S_X$ and $S_X \, \rho \; = \; \rho \, S_X \; = \; \rho^{-1} S_X \; = \; S_X \, \rho^{-1}$.

Proof Since $\rho \in N(S_X)$, we have $\rho \, S_X \, \rho^{-1} = S_X$, therefore also $S_X =$ $\rho^{-1} S_X \, \rho$, and also $\rho \, S_X = S_X \, \rho$. Since $\rho^2 \in S_X$, we find $S_X \rho\rho = S_X$, so $S_X \, \rho = S_X \, \rho^{-1}$ and similarly from $\rho\rho S_X = S_X$ we get $\rho S_X = \rho^{-1} S_X$. This completes the proof.

Let ρ be an involution for the reference isomer in $J_X(L)$. Regarding the conversion of S_X to ρS_X as an isomerization process, an intermediate species can be considered to be a racemic mixture of X and its involution isomer; we take as the ρ-racemate group the chemical identity group of that intermediate species. In precise terms

5.3 Definition If ρ is an involution for X in $J_X(L)$, the ρ-racemate group of X is the Dieter group $D[S_X, \rho S_X]$, i. e. the set of all $\alpha \in \text{Sym}L$ that preserve the chemical identity of the system $\{S_X, \rho S_X\}$.

We can explicitly calculate this racemate group

5.4 Theorem The ρ-racemate group of X is $R = S_X \cup \rho S_X$. In particular, the chemical identity group of X is a normal subgroup of index 2 in the ρ-racemate group of X.

Proof According to the discussion following 4.3, we have

$$D[S_X, \rho S_X] \;=\; \text{permanent} \;\begin{vmatrix} S_X & \rho \, S_X \\ S_X \, \rho^{-1} & \rho \, S_X \, \rho^{-1} \end{vmatrix}$$

$$=\; [S_X \cap \rho \, S_X \, \rho^{-1}] \;\cup\; [\rho \, S_X \cap S_X \, \rho^{-1}]$$

and, by 5.2, this is $S_X \cup \rho S_X$ (see VII,1).

We now proceed in analogy to our development in 2. Let $J_X(L)$ be a family of permutation isomers with reference isomer X having chemical identity group S_X, and let ρ be an involution for X. We start with the reference model E, but this time we take it with the group $R_X = S_X \cup \rho S_X$, instead of the group S_X. The resulting ρ-racemate $\{X,\overline{X}\}$ can be used as a reference racemate with the family of permuted models $P_E(L)$ exactly as the reference isomer X is used. The cosets of R then represent the ρ-racemates within the family of permutation isomers, so we have a family of permuted ρ-racemates $\overline{J_X(L)}$.

As in 2, the decomposition of SymL by the cosets μR_X represents the set of chemically distinct ρ-racemates. Since these isomeric racemates are precisely $\mu R_X = \mu S_X \cup \mu \rho S_X$, we observe that they are simply pairs $\{\mu S_X, \mu \rho S_X\}$ of isomeric ρ-racemates from the original family $J_X(L)$. Thus, calling $\mu \rho S_X$ the ρ-enantiomer of μS_X in $J_X(L)$, the family $\overline{J_X(L)}$ provides a convenient way to discuss the set of pairs consisting of isomers and their ρ-enantiomers, in $J_X(L)$. Moreover, the chemical identity group of $X,\overline{X}$ in $\overline{J_X(L)}$ and that of pairs $\{\mu S_X, \mu \rho S_X\}$ in $J_X(L)$ are related by

5.5 Theorem In $\overline{J_X(L)}$, the chemical identity of the racemate represented by $\{\mu S_X, \mu \rho S_X\}$ is $\mu R_X \mu^{-1}$. This is the same as the Dieter group of the system $\{\mu S_X, \mu \rho S_X\}$ in $J_X(L)$.

Proof The first part is immediate from 3.2. For the second part, the 5.2 Proposition gives

$$D[\mu S_X, \mu\rho S_X] = \text{permanent} \begin{vmatrix} \mu\ S_X\ \mu^{-1} & \mu\ \rho\ S_X\ \mu^{-1} \\ \\ \mu\ S_X\ \rho^{-1}\mu^{-1} & \mu\ \rho\ S_X\ \rho^{-1}\mu^{-1} \end{vmatrix}$$

$$= \mu\ S_X\ \mu^{-1}\ \cup\ \mu\ \rho\ S_X\ \mu^{-1}\ =\ \mu\ R\ \mu^{-1}$$

The permutations in $\mu R\mu^{-1}$ decompose into two disjoint sets, $\mu S_X\mu^{-1}$ and $\mu\rho S_X\ \mu^{-1}$; and for the action of these permutations on the isomers $\mu S_X, \mu\rho S_X$ in $J_X(L)$ we get

5.6 <u>Corollary</u> The permutations $\alpha \in \mu S_X\mu^{-1}$ preserve the chemical identity of μS_X and of its ρ-involution isomer $\mu\rho S_X$. The $\alpha \in \mu\rho S_X\mu^{-1}$ convert each one of $\mu S_X, \mu\rho S_X$, to the other.

<u>Proof</u> We need to verify $\mu S_X\mu^{-1}\mu S_X = \mu S_X$ and $\mu S_X\mu^{-1}\mu\rho S_X = \mu\rho S_X$ for the first part, and this follows from 5.2, as does the verification that $\mu\rho S_X\mu^{-1} = \mu\rho S_X$ and $\mu\rho S_X\mu^{-1}\mu\rho S_X = \mu S_X$ for the second part.

We have seen in 3.3 that there are $[N(S_X):S_X]$ isomers in $J_X(L)$ having the same chemical identity group S_X. To investigate the discrimination ability of the racemate groups, we find as in 3.3 that there will be $[N(R_X):R_X]$ distinct ρ-racemates in $\overline{J_X(L)}$ having the same racemate group, R_X.

Now, although $S_X \subset R_X$, it is not necessarily true that $N(S_X) \subset N(R_X)$ or conversely; however, we do have $R_X \subset N(S_X)$ because S_X is normal in R_X. Thus the situation of these groups can be pictured, in general, as

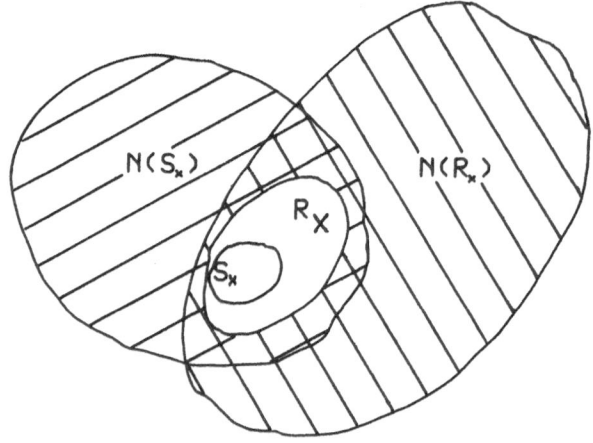

and we consider certain cases.

5.7 Proposition If $\pi \in N(S_X) - N(R_X)$ then

1. In $J_X(L)$ the isomer πS_X is distinct from S_X and its involution ρS_X. But it has the same chemical identity group as X.

2. In $\overline{J_X(L)}$ the racemate group of $\{\pi S_X, \pi\rho S_X\}$ is different from that of $\{S_X, \rho S_X\}$.

Proof Since $\pi \in N(S_X)$, the chemical identity group of πS_X is $\pi S_X \pi^{-1} = S_X$; and since $\pi \notin N(R_X)$ it does not belong to $R_X = S_X \cup \rho S_X$. Thus πS_X is neither S_X nor its enantiomer ρS_X. The racemate group of $\{\pi S_X, \pi\rho S_X\}$ being $\pi R_X \pi^{-1} \neq R_X$ since $\pi \notin N(R_X)$, the proof is complete.

5.8 Proposition If $\pi \in N(R_X) - N(S_X)$ then

1. In $J_X(L)$ the isomer πS_X has a chemical identity group different from S_X, so it is neither the isomer S_X nor its involution ρS_X.

2. In $\overline{J_X(L)}$ the racemate group of $\{\pi S_X, \pi\rho S_X\}$ is the same as that of $\{S_X, \rho S_X\}$.

<u>Proof</u>   Since $\pi \notin N(S_X)$ the chemical identity group of $\pi S_X \neq S_X$; because

$\pi \in N(R_X)$ we find $\{\pi S_X, \pi \rho S_X\}$ has racemate group $\pi R_X \pi^{-1} = R_X$, the same

as that of $\{S_X, \rho S_X\}$.

<u>5.9 Theorem</u>   If $\pi \in [N(R_X) \cap N(S_X)] - R_X$, then

1. In $J_X(L)$ the isomer πS_X is different from S_X and its involution

    isomer ρS_X . But it has the same chemical identity group as X.

2. In $\overline{J_X(L)}$, the racemate group of $\{\pi S_X, \pi \rho S_X\}$ is the same as that of

    $\{S_X, \rho S_X\}$. In particular, exactly the same permutations preserve the

    chemical identity πS_X as of S_X , and exactly the same permuta-

    tions interconvert πS_X to $\pi \rho S_X$ as do S_X to ρS_X .

<u>Proof</u>   Because $\pi \notin R_X = S_X \cup \rho S_X$, we have πS_X is different from S_X and

ρS_X . But $\pi \in N(S_X)$ shows the chemical identity group of πS_X to be S_X and,

because $\pi \in N(R_X)$, we find that $\{\pi S_X, \pi \rho S_X\}$ has the same racemate group

as $\{S_X, \rho S_X\}$.

From 5.7, 5.8 we conclude that some isomers may have the same chem-

ical identity group but different racemate groups, while others may have

different chemical identity groups but   the   same   racemate   groups

(see VII,1.3).   The   5.9 Theorem indicates that there can exist the

phenomenon of hyperchirality [4,6]: chemically distinct species in $J_X(L)$

with the property that exactly the same permutations preserve the chemical

identity of both the isomers and their ρ-racemates, and exactly the same

permutations convert each isomer to its ρ-racemate (see also VII,1.3).

6.0 __Definition__   A   set   of   enantiomer   pairs $\{(\mu_1 S_X, \mu_1 \rho S_X), \ldots,$
$(\mu_n S_X, \mu_n \rho S_X)\}$, $n \geq 2$,   in $J_X(L)$ is called a hyperchiral family if all of
them have exactly the same chemical identity group and exactly the same
ρ-racemate group. Any two non ρ-enantiomeric isomers belonging to a hyperchiral family are called hyperchiral isomers [4].

As the definition indicates, neither their chemical identity groups
nor   their   racemate   groups can distinguish between members of hyperchiral
families.

The number of hyperchiral families in any given family of permutation
isomers is easy to determine.   The   number of such enantiomer pairs is
$[\{N(R_X) \cap N(S_X)\}:R_X]$; and since $[SymL:R_X] = [SymL:\{N(R_X) \cap N(S_X)\}]$
$\cdot[\{N(R_X) \cap N(S_X)\}:R_X]$ there will be $[SymL:\{N(R_X) \cap N(S_X)\}]$ different
hyperchiral families, each containing $[\{N(R_X) \cap N(S_X)\}:R_X]$ pairs.

__Remark__ It is interesting to observe that there may exist a $\pi \in SymL$ with
$\pi \in \{N(R_X) - R_X\} \cap \{N(S_X) - S_X\} = \{N(R_X) \cap N(S_X)\} - R_X$ and $\pi^2 \in R_X$. In
this case, the π is different from ρ, and is an involution in $\overline{J_X(L)}$ so
that it may itself be used to form π-involutions in $\overline{J_X(L)}$ to get a family
$\overline{\overline{J_X(L)}}$.

References

[1] J. Ugi, H. Klusacek, G. Gokel, P. Hoffmann and P. Gillespie, Angew.
     Chem., __82__, 741 (1970); Angew. Chem. Int. Ed., __9__, 703 (1970); see also:
     D. J. Klein and A. H. Cowley, J. Amer. Chem. Soc., __97__, 1633 (1975);
     Hässelbarth and E. Ruch, Isr. J. Chem. __15__, 112 (1977); R. Kopp,
     Dissertation, Techn. Universität München, 1979.

[2] J. Gasteiger, P. D. Gillespie, D. Marquarding and I. Ugi, Topics Curr. Chem., 48, 1 (1974).

[3] The representation of permutational isomers by ligand permutations alone, without any explicit reference to skeletal indexing and to representations of skeletal symmetry [4] is an essential feature in the theory of chemical identity groups, enabling us to avoid the conceptual difficulties which arise from double indexing. In double indexing systems the ligands and the skeleton are indexed independently, or the indices of the skeletal sites are assigned to the ligands which belong to the respective sites [5].

[4] J. Dugundji, D. Marquarding and I. Ugi, Chemica Scripta 9, 74 (1976); 11, 17 (1977).

[5] see e.g.: E. Ruch and A. Schönhofer, Theor. Chim. Acta, 10, 91 (1968).

[6] see also: L. G. Harrison and T. C. Lacalli, Proc. Roy. Soc. London B 202, 361 (1977).

REACTION SCHEMES

Within the present conceptual framework, molecules with some indistinguishable ligands and ligand-preserving isomerization processes can both be treated in a similar manner by using the notion of a reaction scheme.

1.   *Partitions and Coverings in SymL*

Let $J_\chi(L)$ be a family of permutation isomers with all ligands chemically distinguishable, having reference model E and chemical identity group S_χ. The cosets $\{\lambda S_\chi\}$ of S_χ give a well-defined, fixed partition of SymL, with permutations λ,μ belonging to the same coset if and only if λE is chemically equivalent to μE. Now, there are various conditions based on chemical considerations that we can impose on the study of $J_\chi(L)$, e. g. some of the ligands may be stipulated to be chemically indistinguishable; and such conditions may force us to declare the models $\lambda E, \mu E$ to be chemically identical, or similar in some respect, even though λ,μ belong to different cosets. The chemically meaningful situations arise from conditions T that force us to regard all the permutations in distinct cosets $\lambda S_\chi, \mu S_\chi$ as representing equivalent (see V,2) or directly interconvertible (see V,3) chemical species, rather than from conditions forcing the "merger" of only a few selected permutations. In this section, we discuss such a condition abstractly, and in sufficient generality to cover the applications given in the next two sections.

Let $\{S | S \in \Lambda\}$ be a partition of SymL and let $\{T | T \in \mathcal{T}\}$ be a covering of SymL, i. e. a family of subsets, not necessarily pairwise disjoint, whose union is SymL. For each $T \in \mathbf{T}$, let $StT = \{S \in \Lambda \mid S \cap T \neq \emptyset\}$, i. e. the set StT, called the star of T relative to Λ, is the family of all the sets $S \in \Lambda$ that meet T.

Abstractly, the situations considered here start by merging all the sets $S \in \Lambda$ belonging to any single StT; we call this merging a reaction of T on Λ . Since a given $S \in \Lambda$ may belong to more than just one StT, the maximal family of sets to be merged with a given S is generally considerably larger than StT. The problem is to give a direct description of all the maximal families of merging sets.

For this purpose, we will use the sets $T \in \mathbf{T}$ as counters, gradually enlarging each family StT until we get the desired families. Geometrically, the process, called a reaction scheme of T on Λ, can be visualized as follows: On the first of two transparent films of SymL, draw its covering by the sets $T \in \mathbf{T}$, and on the second, its decomposition by the sets $S \in \Lambda$. Lay the S-film on top of the T-film and for each $T \in \mathbf{T}$, mark the sets S that cover it; these are the sets StT. We now enlarge each StT to a set $U_2(T)$ by adding to StT all the StT' it meets; clearly the sets making up $U_2(T)$ all merge. We next enlarge each $U_2(T)$ by adding to it all the StT' it meets, and continue in this way. There being only $|\Lambda|$ sets S available, this enlarging process finally stops; the distinct $U_i(T)$ will be a partition of Λ, and each will be a maximal family of merging sets. It is, of course, quite possible that all the $S \in \Lambda$ merge into one (see VI,4). We now give the formal details of this process.

1.1 **Definition** Let $\{S \mid S \in \Lambda\}$ be a partition, and $\{T \mid T \in \mathbf{T}\}$ a covering, of SymL. By a reaction scheme on Λ by T is meant a sequence $Z_i = \{U_i(T)|T \in T\}$ of coverings of SymL where

$$U_1(T) \quad = \quad StT$$

$$U_2(T) \quad = \quad \cup \ \{StT'|StT' \cap U_1(T) \neq \emptyset\}$$

. . . .

$$U_{n+1}(T) = \quad \cup \ \{StT'|StT' \cap U_n(T) \neq \emptyset\}$$

. . . . .

For a fixed i, the covering Z_i of SymL is called the ith stage of the reaction scheme.

It is useful to have a direct characterization of the sets $S \in \Lambda$ that belong to a given $U_i(T)$. To get this, call chain from $T_1 \in T$ to $S_n \in \Lambda$ any sequence $T_1, S_1, T_2, S_2, \ldots, T_n, S_n$ (alternating between $T_i \in T$ and $S_i \in \Lambda$) in which the intersection of each two adjacent terms is nonempty (i. e. $T_1 \cap S_1 \neq \emptyset$, $S_1 \cap T_2 \neq \emptyset, \ldots, T_n \cap S_n \neq \emptyset$) (see VI,4 Table 5). We now have the simple

1.2   Proposition A set S belongs to $U_i(T)$ for some i, if and only if there is a chain joining T to S.

Proof   Only if: Observe that if each $S \in U_i(T)$ is chained to T, so also is each $S' \in U_{i+1}(T)$: for $S' \in StT'$ where $StT' \cap U_i(T) \neq \emptyset$; this intersection contains an $S \in U_i(T)$ which is chained to T; adding T', S' at the end of that chain gives a chain from T to S'. Since each $S \in StT = U_i(T)$ is chained to T, the result follows by induction.

If: Let $T, S', T', S'', T'', \ldots, S$ be a chain from T to S. From the portion

T,S',T',S" we find S' ∈ StT ∩ StT' = U₁(T) ∩ StT' so the intersection is nonempty, therefore S" ∈ StT' ⊂ U₂(T). From this and T',S",T",S"', we find S" ∈ StT" ∩ U₂(T) is nonempty, therefore S"'∈ StT" ⊂ U₃(T). Proceeding recursively, this shows that S is in U₁(T) for some i.

The main result on reaction schemes is

1.3   Theorem   Given a reaction scheme of T on Λ, then from some stage i ≤ |Λ| onward,

1. $U_i(T) = U_{i+s}(T)$ for each T ∈ T and all s ≥ 0

2.   The distinct sets $U_i(T)$ in $\{U_i(T)|T ∈ T\}$ will be a partition of Λ; each such $U_i(T)$ is a maximal family of merged sets.

Proof   Fix T ∈ T. We have $U_1(T) ⊂ U_2(T) ⊂ ...$; since there are only |Λ| sets S available, and since each increase adds at least one set S, there must be a first time j = j(T), with j ≤ |Λ| that $U_j(T) = U_{j+1}(T)$. This condition is equivalent to the statement that each StT' meeting $U_j(T)$ is contained in $U_j(T)$, which implies that no additional sets will be added to $U_j(T)$: it is a maximal family of merging sets. This result is true for each T ∈ T; letting i = max[j(T)|T ∈ T] we have i ≤|Λ| and $U_{i+s}(T) = U_i(T)$ for all T ∈ T and all s ≥ 0, thereby proving (1). To prove (2), it is enough to show that any two families $U_i(T)$, $U_i(T')$ are either identical or disjoint. Suppose $U_i(T) ∩ U_i(T') ≠ ∅$; they have a set S ∈ Λ in common that can be reached by chains T,...,S and T',...,S; we therefore have a chain from T to T'. Now let Ŝ ∈ $U_i(T)$ then Ŝ is chained to T, which chains to T', so Ŝ ∈ $U_k(T')$ for some k; since all $U_k(T') ⊂ U_i(T')$ by our choice of i, this shows Ŝ ∈ $U_i(T')$ and, since Ŝ ∈ $U_i(T)$ is arbitrary, that

$U_i(T) \subset U_i(T')$. Similarly, $U_i(T') \subset U_i(T)$ so $U_i(T) = U_i(T')$ and the proof is complete.

The sets $S \in \Lambda$ contained in any one of the $U_i(T)$ that partition Λ are called a T-merging family; the set of permutations $\overline{U_i(T)}$ belonging to the members of $U_i(T)$ is called a T-merging class of permutations. Since the S are pairwise disjoint, the partition of Λ by the T-merging families $U_i(T)$ gives a partition of SymL by the T-merging classes $\overline{U_i(T)}$ of permutations. Thus, we can work directly with permutations (rather than with the $S \in \Lambda$); and we need only calculate the stages of the reaction scheme of T on Λ until the permutations lying in the distinct sets $\{\overline{U_i(T)}|T \in T\}$ form a partition of SymL.

In several cases the Λ and T are both partitions of SymL by cosets of subgroups S_X and Σ; in this case, the construction 1.1 is called simply a reaction scheme of Σ on S_X. In such cases, it will be seen that whenever the cosets are of opposite type (e. g. left cosets of S_X, right cosets of Σ) then each Σ-merging class will be the set of cosets of S_X that make up a (Σ,S_X)-double coset; whenever they are of the same type (e. g. left cosets of S_X and of Σ), then each Σ-merging class consists of the cosets of S_X contained in a single left coset of the subgroup $\langle \Sigma,S_X \rangle$ generated by Σ and S_X.

The set valued mappings which we have also used to represent reaction schemes [1,2] are a special case of the schemes which are presented here.

2. *Ligand Substitutions; Chemically Equivalent Ligands*

Let $J_X(L)$ be a family of permutation isomers with a set of chemically distinguishable ligands $L = \{1,2,\ldots,n\}$, reference model E, and chemical identity group S_X. Let $L' = \{l_1,\ldots,l_n\}$ be another set of ligands, not necessarily all chemically distinguishable, and denote in this case by $[\lambda E]$ the species with each ligand i replaced by l_i. We wish to determine the number of distinct species $[\lambda E]$ as λ runs through SymL.

It is convenient to collect the permutations $\lambda \in$ SymL which will give identical species $[\lambda E]$. For this, define $\lambda \approx \mu$ if $[\lambda E] = [\mu E]$; this is clearly an equivalence relation in SymL; let $\{T | T \in \mathcal{T}\}$ be the equivalence classes. Then for each T all the permutations $\lambda \in T$ give a species $[\lambda E]$, so there are exactly $|\mathcal{T}|$ distinct equivalence classes $[\lambda E]$ as λ runs through SymL.

We have exactly two means at our disposal for showing that two species $[\lambda E]$, $[\mu E]$ are chemically identical: In addition to the condition $[\lambda E] = [\mu E]$ we have one other, based on the commonly accepted chemical principle that, if λE is chemically equivalent to μE when all ligands are distinguishable, then $[\lambda E]$ is chemically equivalent to $[\mu E]$.

These two conditions force us to regard all the cosets λS_X that meet a given $T \in \mathcal{T}$ as representing chemically equivalent species $[\mu E]$. For, suppose $T \cap \lambda S_X \neq \emptyset$; let $\eta \in T$, let $\mu \in \lambda S_X$, and let $\xi \in T \cap \lambda S_X$; then because $\eta, \xi \in T$ we have $[\eta E] = [\xi E]$, and because $\xi, \mu \in \lambda S_X$ we have ξE chemically equivalent to μE so, using the chemical principle stated above, we find $[\eta E]$ is chemically equivalent to $[\mu E]$. This is exactly the situation we have discussed in the previous section: we are dealing with a reaction of T on the partition $\{\lambda S_\chi\}$ of SymL by the cosets of S_χ. The T-merging families therefore all represent a single chemical species $[\lambda E]$. On the other hand, if a coset λS_χ does not belong to a T-merging family A, then no StT containing λS_χ meets it, so λS_χ cannot give any species $[\lambda E]$ chemically equivalent to any of those determined by A. Thus, the chemically distinct isomers when the ligand set L' is used are in 1-1 correspondence with the distinct T-merging families; and to explicitly determine these families, we need only use the reaction scheme of T on the partition $\{\lambda S_\chi\}$.

Before doing so, we get a more detailed description of T: in fact, the decomposition $\{T | T \in T\}$ is precisely the family of right cosets of a subgroup $\Sigma \subset$ SymL. To see this, let the ligand set L' be represented as a pairwise disjoint union $L' = I_1' \cup I_2' \cup \ldots \cup \ldots I_m'$, where the ligands in each I_i' are chemically indistinguishable from one another, but chemically distinguishable from those in any $I_j' \neq I_i'$ (so that L' has m chemically distinct types of ligands). Replacing each l_i by the ligand i gives us a decomposition $I_1 \cup I_2 \cup \ldots \cup I_m$ of L.

Let $\Sigma = \{\sigma \in$ SymL$| \sigma(I_1, \ldots, I_n) = (I_1, \ldots I_n)\}$ be the set of all permutations that map each I_i onto itself (i. e. σ permutes only chemically identical ligands among themselves). It is obvious that Σ is a subgroup of SymL; we call Σ the stabilizer of the ligand substitution $i \mapsto l_i$. Its basic property is the simple

2.1 <u>Lemma</u>   For any $\lambda, \mu \in$ SymL, we have $[\lambda E] = [\mu E]$ if and only if λ, μ belong to a common right coset $\Sigma \eta$ of Σ in SymL.

Proof Assume $\lambda = \sigma\mu$ for some $\sigma \epsilon \Sigma$. Given μE, the exchange $\sigma \epsilon \Sigma$ permutes ligands i that are to be replaced by chemically equivalent l_i, so $[\mu E]$ is the same species as $[\sigma\mu E] = [\lambda E]$. Conversely, if $[\lambda E] = [\mu E]$, then chemically identical ligands are located at each site; the permutation λ therefore can then be obtained from μ by a permutation belonging to Σ, i. e. $\lambda = \sigma\mu$.

With these preliminaries, we now enter into a reaction scheme of Σ on S_X in order to calculate the Σ-merging families.

2.2 Theorem Let $J_X(L)$ be a family of permutation isomers, with all ligands chemically distinguishable, and with chemical identity group S_X. Let $\Sigma \subset SymL$ be the stabilizer of a ligand substitution. Then

1. The Σ-merging families will be sets of cosets λS_X making up a single double coset $\Sigma\eta S_X$ (so that $[\lambda E],[\mu E]$ will be chemically equivalent if and only if λ,μ belong to the same double coset $\Sigma\eta S_X$).

2. Each double coset $\Sigma\lambda S_X$ is formed as the union of

$$\frac{|\Sigma|}{|\lambda S_X \cap \Sigma\lambda|} \quad \text{cosets } \eta S_X$$

   (so that the Σ-merging families are not all equally large).

3. If $\lambda_1,..,\lambda_N$ is a transversal of Σ in $SymL$, where $N = [SymL:\Sigma]$, then the number of chemically distinct species $[\lambda E]$ as λ runs through $SymL$ is

$$\frac{1}{|S_X|} \cdot \sum_{i=1}^{N} |\lambda_i S_X \cap \Sigma\lambda_i|$$

Proof (1) We begin by calculating the terms in the reaction scheme of Σ on S_X; for convenience, we use the underlying sets $\overline{U}_i$ of permutations. We have

$$\overline{U_1(\Sigma\lambda)} \;=\; \cup\;\{\mu S_\chi \mid \mu S_\chi \cap \Sigma\lambda \neq \emptyset\}$$

$$=\; \cup\;\{\mu S_\chi \mid \mu \in \Sigma\lambda\} \;=\; \Sigma\lambda S_\chi$$

since $\mu S_\chi \cap \Sigma\lambda \neq \emptyset$ means μS_χ contains an element $\gamma \in \Sigma\lambda$, so $\mu S_\chi = \gamma S_\chi$.

The double cosets being a partition of SymL, the distinct double cosets $\Sigma\lambda S_\chi$ give us the Σ-merging classes of permutations.

(2) Given any λS_χ, we want to find the number of cosets $\sigma\lambda S_\chi$ as σ runs over Σ. Now $\tilde{\sigma}\lambda S_\chi = \sigma\lambda S_\chi$ if and only if $\lambda^{-1}\sigma^{-1}\tilde{\sigma}\lambda \in S_\chi$, i. e. $\sigma^{-1}\tilde{\sigma} \in \lambda S_\chi \lambda^{-1}$; since $\sigma,\tilde{\sigma} \in \Sigma$, this occurs if and only if $\sigma,\tilde{\sigma}$ are in a common coset of $\Sigma \cap \lambda S_\chi \lambda^{-1}$ in Σ, so the number of cosets μS_χ in $\Sigma\lambda S_\chi$ is

$$[\Sigma : \Sigma \cap \lambda S_\chi \lambda^{-1}] \;=\; \frac{|\Sigma|}{|\Sigma \cap \lambda S_\chi \lambda^{-1}|}\;.$$

By considering the bijection $f: \mathrm{SymL} \to \mathrm{SymL}$ given by $x \mapsto x\cdot\lambda$, we have $f(\Sigma \cap \lambda S_\chi \lambda^{-1}) = \Sigma\lambda \cap \lambda S_\chi$, and this completes the proof of (2). The assertion (3) follows from the Burnside-Frobenius theorem, and is proved in the Appendix (8.4).

Polyà [3] was the first to develop a formalism for enumerating the number of isomers when the ligands are not all chemically distinguishable, and that formalism was subsequently generalized by de Bruijn [4]; their method of counting is quite general, and has been used, by Polyà, de Bruijn and others in situations having no connection with chemistry. Subsequently, Ruch et al. [5] observed that the chemically identical molecules are determined by permutations belonging to a single 'double coset, and that the simpler Burnside-Frobenius formula can be used to find the number of chemically distinguishable isomers. Since the notion of a family of permutation isomers [6] was not known at the time, neither Polyà nor Ruch indicated that they were working within a single family of permutation isomers, so some confusion about what exactly is being counted has appeared in the literature. Our approach differs from the previous ones in that it essentially reduces the problem to calculating the number of equivalence classes in the transitive closure of a reflexive, symmetric relation.

The formalisms of Polyà and of Ruch do not readily answer the question of which isomers λS_X are converted to the same isomer after a ligand substitution, a matter that is clear in our approach. Moreover, when we start with a chiral family $J_X(L)$, we can determine which of the resulting isomers is chiral. For this we need

2.3 Lemma Assume X is a chiral molecule with an achiral skeleton, so that it has an enantiomer in the same family of permutation isomers. Let $\Sigma \lambda S_X$ be a double coset. If it contains the enantiomer coset of any one $\mu S_X \subset \Sigma \lambda S_X$, then it contains the enantiomer coset of every $\mu S_X \subset \Sigma \lambda S_X$.

Proof Let μS_X and its enantiomer coset $\mu \rho S_X$ belong to the double coset $\Sigma \lambda S_X$. We can take μ to represent the double coset, so $\Sigma \lambda S_X = \Sigma \mu S_X$. Now let $\sigma \mu S_X$ be any coset in $\Sigma \lambda S_X$; we are to show $\sigma \mu \rho S_X \subset \Sigma \lambda S_X$. But, since $\mu \rho S_X \subset \Sigma \lambda S_X$, also $\sigma \mu \rho S_X \subset \sigma \Sigma \lambda S_X$, and the proof is complete.

This leads to

2.4 Theorem Assume $J_X(L)$ is chiral, with chemical identity group S_X. Let Σ be the stabilizer of a ligand substitution. The isomer represented by the double coset $\Sigma \lambda S_X$ will be achiral if and only if the double coset contains the enantiomer coset of λS_χ.

<u>Proof</u> By 2.3, the double coset $\Sigma\lambda S_\chi$ contains either all, or none, of the enantiomers of its member cosets. In the first case, it is achiral; in the second case it is chiral, with the isomer represented by $\Sigma\lambda\rho S_\chi$ (ρ the enantiomerization) being the enantiomer of $\Sigma\lambda S_\chi$.

Using the enantiomerization ρ, the result in 2.4 can be stated directly in terms of permutations: the isomer represented by $\Sigma\lambda S_\chi$ will be achiral if and only if the enantiomerization $\rho \in \lambda^{-1}\Sigma\lambda$: for, the enantiomer coset $\lambda\rho S_\chi \subset \Sigma\lambda S_\chi$ if and only if $\lambda\rho \in \Sigma\lambda$ (see VII,1.1h).

3. *Ligand-preserving Isomerizations and Reaction Schemes*

Although we have already discussed isomerization mechanisms (see II,5 and V,4), we have not considered how the various isomers are rearranged in such a process. In this section, we give a simple technique for tracing the dynamically and experimentally observable aspects in the formation of isomers of $J_\chi(L)$, e. g. which isomers can be formed with a given isomerization process, which isomers are directly connected, and which are connected via certain intermediates. We shall illustrate the technique in a simple instance, and then indicate how it can be used to handle more general cases (see VII,2).

Let $J_\chi(L)$, $J_Y(L)$ be two families of permutation isomers with the same set L of chemically distinguishable ligands, reference models E,E' and chemical identity groups S_χ,S_Y. By an isomerization process $J_\chi \rightleftharpoons J_Y$

is meant the conversion of each molecule μE to $\mu E'$. In practice, the pro-

cess is represented by a diagram, such as

called the reference model of the isomerization, which indicates how the

ligands on E will be distributed on E'; the two models of this diagram are

taken to be the reference models of their respective families.

We now study the evolution of the isomers in $J_X(L)$. We have two

partitions of SymL, by the cosets of S_X and those of S_Y. Observe now that

if any two cosets $\mu S_X, \lambda S_X$ meet a single ηS_Y, then $\mu S_X, \lambda S_X$ can be inter-

converted by this isomerization: for there is a molecule in μS_X going to

ηS_Y, so that the isomer ηS_Y is formed, and there is a molecule of ηS_Y that

goes to λS_X, so that λS_X is reached.

This means that we want to regard as similar (i. e. interconvertible

by the isomerization) all the isomers λS_X that meet a single coset ηS_Y.

This is precisely the situation considered in section 1: we have therefore

a reaction of S_X on S_Y. To determine the maximal classes of interconvert-

ing isomers of $J_X(L)$, we therefore need only calculate the final terms in

the reaction of S_X to S_Y.

3.1 Theorem Let $J_X(L) \rightleftharpoons J_Y(L)$ be an isomerization process. Then

1. The set of isomers of $J_X(L)$ that will be reached from a single

   isomer λS_X is the set of all isomers μS_X that are contained in

   the left coset $\lambda \langle S_X, S_Y \rangle$ of the group generated by S_X and S_Y.

2. Each coset λS_X will be directly connected with

$\dfrac{|S_Y|}{|S_X \cap S_Y|}$ other cosets (so that each merging class has the

same number of cosets).

<u>Proof</u> For the first stage of the reaction of S_X to S_Y we get

$$U_1[\mu S_Y] = \{\lambda S_X \mid \lambda S_X \cap \mu S_Y \neq \emptyset\}$$

$$= \{\lambda S_X \mid \lambda \in \mu S_Y\}$$

$$= \mu S_Y S_X$$

Since $S_Y S_X$ is not, in general, a group (unless $S_X S_Y = S_Y S_X$) this is not,

in general, a partition of SymL; so we proceed to the second stage, and,

in order to simplify the notation, we will write S instead of S_X and T

instead of S_Y. Now

$$U_2[\mu \ T] = \{\lambda TS \mid \lambda TS \cap \mu TS \neq \emptyset\}.$$

We show $\lambda TS \cap \mu TS \neq \emptyset$ if and only if $\lambda \in \mu TST$. For, if $\lambda ts = \mu t_o s_o$ then

$\lambda = \mu t_o s_o s^{-1} t^{-1} \in \mu TST$; conversely, if $\lambda \in \mu TST$, then $\lambda = \mu t s_o t_o$ so

$\lambda t_o^{-1} = \mu t s_o$ and therefore $\lambda t_o^{-1} S = \mu t S$; but this says $\lambda TS \cap \mu TS \neq \emptyset$. Thus,

$$U_2[\mu T] = \{\lambda TS \mid \lambda \in \mu TST\} = \mu TSTTS = \mu TSTS$$

Similarly,

$$U_3[\mu T] = \mu TSTSTS$$

and so on. We know that this process of expansion will stop at a certain

$i \leq [\text{SymL}:S_X]$. It is easy to see (e. g. from the fact that $U_i(T) = U_{2i}(T)$)

that the product of i terms TS...TS is then a subgroup of SymL, and in

fact, the subgroup $\langle T,S \rangle$ generated by T and S. Thus, the T-merging

families form the cosets of $\langle T,S \rangle$; the coset λS_X will encounter the coset

μS_X if and only if both of these cosets are contained in a single left coset of $\langle S_X, S_Y \rangle$. In particular, there are $[\text{SymL}: \langle S_X, S_Y \rangle]$ distinct families of interconverting isomers.

For (2) we want to find the number of distinct cosets of S_X in the star of λS_Y, i. e. the number of distinct cosets $\lambda t S_X$ as t runs over S_Y. Now $\lambda t S_X = \lambda \hat{t} S_X$ if and only if $(\lambda t)^{-1}(\lambda \hat{t}) = t^{-1} \hat{t} \in S_X$; since $t^{-1}, \hat{t} \in S_Y$, this says $t^{-1} \hat{t} \in S_X \cap S_Y$, so that $\tilde{t} \in t(S_X \cap S_Y)$: the $t, \hat{t}$ must be in the same coset of S_Y by $S_X \cap S_Y$. The number of such cosets being $|S_Y| / |S_X \cap S_Y|$, the proof is complete.

Recalling 1.2 Proposition, the interconversion can be illustrated graphically. Represent the S_X cosets by a row of points $a_1, \ldots, a_s$ and the S_Y cosets by a parallel row of points, $b_1, \ldots, b_t$. For each S_X coset a_i, draw a line to each b_k that it meets. Then, by reading first from any a_i, to the points $\{b_j\}$ joined to it, and from each one of those b_j back to the points $\{a_l\}$ joined to them, and repeating this process, we can trace the evolution of any given isomer a_i. Note that there is no need to construct separately the lines joining the b_j to the a_i, since if b_j meets a_i, the link between a_i and b_j has already been drawn. With this diagram, the question whether an isomer can be converted to another by the given iso-merization can be easily answered, and indeed all conceivable pathways by which this can be done are found.

The process can be equally well applied in isomerizations A $\rightleftharpoons$ B when the ligands are not all chemically distinguishable: If Σ is the stabi-lizer of the ligand substitution, we need only find the final stage of the reaction of the covering $\{\Sigma \lambda S_B | \lambda \in \text{SymL}\}$ on the partition $\{\Sigma \lambda S_A | \lambda \in \text{SymL}\}$.

A similar technique can be used for isomerizations A ⇌ B ⇌ C. To find the interconverting isomers in A, one need only construct the first stage {F} of the reaction of S_C on S_B, and then find the reaction of {F} on S_A. Since {F} is in general a covering of SymL rather than a partition, it is in order to handle these general cases that the covering {T | T ∈ T} of 1.1 Definition was not required to be a partition. More complicated cases can be treated in the same way.

4. *Musher Modes and Permutational Isomerizations*

Let $J_X(L)$ be a family of permutation isomers having the reference model E, the chemical identity group S_X, the racemate group R_X and an enantiomerization ρ.

Given any permutation $\mu \in$ SymL, we can regard μ as a motion of E. Then, the motion μ applied to any model λE yields the model $\lambda\mu E$, i. e. first apply the motion μ to E and then make the required ligand interchange λ. The motion $\rho^{-1}\mu\rho$ can be regarded as a "mirror image motion": if applied to the enantiomer ρE of E, it gives $\rho\rho^{-1}\mu\rho E = \mu\rho E$, the enantiomer of μE.

<u>4.1 Definition</u> The set of all isomers obtained from the reference isomer by applying the motions μ and $\rho^{-1}\mu\rho$, is called the Musher mode $M[\mu]$ of the motion μ [7-10][*].

---

[*] The classification of isomerizations corresponding to the Musher modes was first introduced by Gielen et al. [7a].

4.2 **Theorem** The Musher mode M[μ] is the family of all cosets in the union

of the (S_X,S_X) double coset containing μ and that containing $\rho^{-1}\mu\rho$, i. e.

$$M[\mu] = S_X\mu S_X \cup S_X\rho^{-1}\mu\rho \, S_X$$

Proof The reference isomer is represented by $\{sE \mid s \in S_X\}$; the motion μ

applied to these models gives us the models sμE representing the cosets

$s\mu S_X$. Since $s \in S_X$ is arbitrary, we get all the cosets belonging to the

double coset $S_X\mu S_X$. Applying the motion $\rho^{-1}\mu\rho$ to all the members of

$\{sE \mid s \in S_X\}$ gives, in the same way as before, the double coset

$S_X\rho^{-1}\mu\rho S_X$.

Note that because ρ is an enantiomerism, $S_X\rho^{-1} = \rho^{-1}S_X$ and $\rho S_X = S_X\rho$;

thus, the Musher mode of μ can be regarded as the set of all permutations

belonging to $S_X\mu S_X \cup \rho^{-1}(S_X\mu S_X)\rho$, i. e. M[μ] is the set of all per-

mutations in the set $S_X\mu S_X$ and its conjugate by ρ.

There is another way of looking at Musher modes. Given any motion μ,

we say that the type of that motion relative to the racemate group R_X is

the Wigner subclass (see Appendix, 4) $W_R[\mu] = \{r^{-1}\mu r \mid r \in R_X\}$. Since any

two Wigner subclasses are identical or disjoint, and the distinct sets

$\{W_R[\mu] \mid \mu \in SymL\}$ partition SymL, all the permutations of SymL are

divided into mutually distinct types relative to R_X. The permutations in

any one class $W_R[\mu]$ are called "symmetry equivalent relative to R_X" [6,10].

We now have two partitions of SymL: that by the cosets $\{\lambda S_X\}$, and that

by the Wigner subclasses $\{W_R[\mu]\}$, so we can speak of a reaction of $\{W_R[\mu]\}$

on $\{\lambda \, S_X\}$; the Musher mode of μ is the first stage of this reaction:

4.3   Corollary   The Musher mode $M[\mu] = StW_R[\mu]$, i. e. it is the set of all cosets λS_X that meet $W_R[\mu]$.

Proof   We have

$$StW_R[\mu] = \{\lambda S_X \mid \lambda S_X \cap W_R[\mu] \neq \emptyset\}$$

Now, this intersection will be non-empty if and only if λ is such that $\lambda s = r^{-1}\mu r$ for some $s \in S_X, r \in R_X$, i. e. for all λ of the form $r^{-1}\mu rs$ for some $r \in R_X$, $s \in S_X$, so

$$StW_R[\mu] = \{r^{-1}\mu rsS_X \mid s \in S_X, r \in R_X\}$$

Since $R_X = S_X \cup \rho\, S_X$ is a disjoint union, we have $r \in S_X$ or $r \in \rho S_X$. In the first case we get $\{r^{-1}\mu rsS_X \mid r \in S_X\} = S_X\mu S_X$. In the second case, we get

$$\{s_1^{-1}\rho^{-1}\mu\rho s_1 sS_X\} = \{S_X\rho^{-1}\mu\rho\ S_X\}$$

so

$$StW_R[\mu] = S_X\mu S_X \cup S_X\rho^{-1}\mu\rho S_X = M[\mu]$$

and this completes the proof.

We can also speak of a reaction of the Wigner subclass $W_R[\mu]$ on the partition of SymL by the family of racemate cosets $\{\lambda R_X\}$. For this case,

$$StW_R[\mu] = R_X\mu R_X$$

and since $R_X = S_X \cup \rho S_X$, a disjoint union, this leads to

$$StW_R[\mu] = S_X\mu S_X \cup \rho(S_X\mu S_X) \cup (S_X\mu S_X)\rho \cup \rho^{-1}(S_X\mu S_X)$$

$$= M[\mu] \cup \rho(S_X\mu S_X) \cup (S_X\mu\ S_X)\rho$$

$$= M[\mu] \cup S_X\rho\mu S_X \cup S_X\mu\rho S_X.$$

References

[1] R. Kopp, Dissertation, Techn. Universität München 1979.

[2] J. Dugundji, J. Showell, R. Kopp, D. Marquarding and I. Ugi, Isr. J. Chem. 20, 20 (1980).

[3] G. Polya, Acta Math. 68, 145 (1937); see also: S. W. Colomb, "Information Theory", The Universities Press, Belfort 1961; F. Harary, E. M. Palmer, R. W. Robinson and R. C. Read, in: "Chemical Applications of Graph Theory", A. T. Balaban, ed., Academic Press, London 1976, p. 11.

[4] N. G. De Bruijn, Koninkl. Ned. Akad. Wetenschap. Proc. Ser. A62, 59 (1959); Niew Arch. Wiskunde (3) 18, 61 (1970).

[5] E. Ruch, W. Hässelbarth and B. Richter, Theoret. Chim. Acta (Berl.) 19, 288 (1970); W. Hässelbarth and E. Ruch, ibid. 29, 259 (1973).

[6] I. Ugi, D. Marquarding, H. Klusacek, G. Gokel and P. Gillespie, Angew. Chem. 82, 741 (1970); Angew. Chem. Int. Ed. 9, 703 (1970).

[7] a) M. Gielen, J. Brocas, M. De Clerq and G. Mayence, Proc. of the 3. Symp. Coord. Chem., Vol. 1, Ed. M. T. Beck, Brussels 1970, p. 495; M. Gielen and N. van Lautem, Bull. Soc. Chim. Belges 79, 679 (1970); 80, 207 (1971); b) J. I. Musher, J. Amer. Chem. Soc. 94 5662 (1972); Inorg. Chem. 11, 2335 (1972); J. Chem. Educ. 51, 94 (1974); J. Brocas, Top. Curr. Chem. 32, 44 (1972); J. Brocas and R. Willem, Bull. Soc. Chim. Belges 82, 469, 629 (1973);
D. J. Klein and A. H. Cowley, J. Amer. Chem. Soc. 97, 1633 (1975); J. G. Nourse, ibid. 99, 2063 (1977); see also: J. Brocas, M. Gielen and R. Willem, "The Permutational Approach to Dynamic Stereochemistry", McGraw-Hill, New York 1983, Ch. 4, 9-12.

[8] W. G. Klemperer, J. Chem. Phys. 56, 5478 (1972); J. Amer. Chem. Soc. 94, 6940, 8360 (1972); 95, 380, 2105 (1972); Inorg. Chem. 11, 2668 (1972).

[9] W. Hässelbarth and E. Ruch, Theoret. Chim. Acta, 29, 259 (1973).

[10] J. Dugundji, P. Gillespie, D. Marquarding, I. Ugi and F. Ramirez in: "Chemical Applications of Graph Theory", A. T. Balaban, ed., Academic Press, London 1976, p. 107.

[11] E. P. Wigner, "Spectroscopic and Group Theoretical Methods in Physics (Racah Mem. Vol.), North Holland Publ. Co. Amsterdam 1971, p. 131; Proc. Roy. Soc. (London) A322, 181 (1971).

STRUCTURE OF THE CHEMICAL IDENTITY GROUP

In this Chapter we study the structure of the chemical identity group. This will enable us to choose subgroups that have clear-cut geometrical/ chemical meanings, and to express the chemical identity group as a semi-direct product of such subgroups, thereby simplifying its construction and use in calculations.

This analysis leads also to several new groups useful in stereochemistry. One is the group of constitution-preserving ligand permutations [1], which permits the determination of the number of distinct stereoisomeric permutation isomers that a given compound has. Another is the flexibility group of a conformationally flexible molecule, enabling us to express the chemical identity group of such a molecule in terms of ligand permutations representing two well-defined types of geometrical motions [2].

THE PATTERN OF A MOLECULE AND ITS ENVELOPING GROUP

Let X be a molecule conceptually dissected into a skeleton and a set of ligands appropriate for the experiment under consideration [3]. We call each skeletal atom of X that has ligands attached to it, a monocentric site. Clearly, the notion of a monocentric site depends on the dissection made, and a given skeletal atom may, or may not, carry ligands.

If X has n monocentric sites $i = 1,\ldots,n$ with each site i having the set $\Omega_i = \{1_{i_1},\ldots,1_{i_{s_i}}\}$ of attached ligands, we call the n-tuple $\{\Omega_1,\ldots,\Omega_n\}$ the pattern of X. The set of all ligands in this pattern is $\Omega = \bigcup_{i=1}^{n} \Omega_i$.

For ease of description, we call the set Ω_i of ligands the fan at the monocentric site i and the ligands $l_{i_1}, \ldots, l_{i_{s_i}}$ the blades of the fan Ω_i.

1. The Enveloping Group

Before introducing any chemical considerations, it is convenient to construct a subgroup of $Sym(\Omega)$ in which the chemical identity group of any molecule with the pattern $\{\Omega_1, \ldots, \Omega_n\}$ will be found.

Let $N = \{h \in Sym\Omega \mid$ for each $i = 1, \ldots, n$ there is a j such

$$\text{that } h(\Omega_i) = \Omega_j\}$$

i. e. N is the set of all permutations of Ω that interchange fans between monocentric sites having the same coordination number.

Because each $h : \Omega \rightarrow \Omega$ is bijective, the condition implies, among other things, that for each i the permutation $h|\Omega_i$ maps the fan Ω_i onto some fan Ω_j with $|\Omega_i| = |\Omega_j|$, and that no two distinct Ω_i are mapped into the same Ω_j. This set of permutations N is in fact a group, because the composition of any two members of N belongs to N. We call N the enveloping group of the pattern $\{\Omega_1, \ldots, \Omega_n\}$; it is a generalized wreath product (see Appendix,9).

We next analyze the structure of the enveloping group. Let

$$A = \{\alpha \in Sym\Omega \mid \alpha(\Omega_i) = \Omega_i \text{ for each } i\}$$

that is, each $\alpha \in A$ permutes only the blades at each monocentric site. This is a subgroup of $Sym\Omega$ because the composition of any two such permutations has the same property. It is shown in 9 of the Appendix that

　　　1.　A is a normal subgroup of N

in fact　2.　N is the normalizer of A in $Sym\Omega$

3. If each $|\Omega_i| \geq 2$, then N is its own normalizer in SymΩ.

Moreover 4. N is the semidirect product A $\wedge$ Q, where Q is a subgroup isomorphic to N/A.

A more direct description of Q can be obtained by starting with the
observation that two elements of N belong to the same coset of A if and
only if they give identical permutations of the monocentric sites. Thus,
partition the set {1,...,n} of monocentric sites into blocks $B_1,...,B_k$ by
placing in the same block all those indices that have the same number of
blades in the fans attached to them. Then N/A is isomorphic to G =
{g $\in$ Sym{1,...,n}|g(B_i) = B_i for all i}. Now order Ω in any way and for
each g $\in$ G   choose λ(g):$\Omega \rightarrow \Omega$ to be the unique order-preserving map that
sends each Ω_i onto $\Omega_{g(i)}$. The set {λ(g)|g $\in$ G} forms a subgroup Q of N
that is isomorphic to G and N = A $\wedge$ Q.

We note that the enveloping group of the pattern {$\Omega_1,...,\Omega_n$} is quite
large, containing $|\Omega_1|!...|\Omega_n|!|B_1|!...|B_k|!$ elements. Furthermore, A
itself can be decomposed into a direct product, as can Q.

2. Decompositions of the Chemical Identity and Racemate Groups

Consider now a molecule X having the pattern {$\Omega_1,...,\Omega_n$} and chemical
identity group S_X. Because S_X preserves the chemical identity of X, the
permutations $\lambda \in S_X$ are made up by permuting the blades at the fans, and
by permuting the fans themselves between monocentric sites having the same
coordination number. Thus, S_X is always contained in the enveloping group
of the pattern {$\Omega_1,...,\Omega_n$}.

The permutations in A $\subset$ N will all preserve the chemical constitution of the molecule X, precisely because they do not change the connectivity list of the molecule; however, they may not all belong to the chemical identity group of X: for example, at a monocentric site i

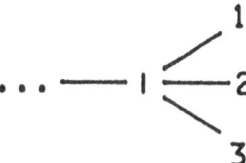

the blade permutation (123) may preserve the chemical identity of X, whereas the blade permutation (12) may not. Thus

$A \cap S_X$ = all ligand permutations of X that preserve its
chemical identity and do <u>not</u> exchange any fans.

These correspond to intraskeletal motions which in the idealized case bring the skeleton into self-coincidence without changing the position of any skeletal atom. We note that because A is normal in N, it follows that $A \cap S_X$ is a normal subgroup of S_X; in many cases $A \cap S_X$ is also abelian (because the blade permutations of each site i form, in general, a cyclic group).

We now observe that $S_X \subset (A \cap S_X) \cdot Q$ because any ligand permutation preserving the chemical identity of X which does not belong to $A \cap S_X$ must exchange fans between distinct sites; it can therefore be expressed as a redistribution of the fans after the blades of each have been permuted in a way that preserves the chemical identity of X.

Since $(A \cap S_X) \subset S_X \subset (A \cap S_X) \cdot Q$ we find from Dedekind's rule that $S_X = (A \cap S_X) \cdot (Q \cap S_X)$, the product being semidirect because $A \cap S_X$ is normal in S_X and $(A \cap S_X) \cap (Q \cap S_X) \subset A \cap Q = \{e\}$. Thus

2.1 <u>Theorem</u> Let S_X be the chemical identity group of a molecule X having the pattern $\{\Omega_1, \ldots, \Omega_n\}$. Then S_X splits into a semi-direct product $S_X = (A \cap S_X) \wedge (Q \cap S_X)$.

The group $Q \cap S_X$ can be intuitively regarded as the fan exchanges that occur in the chemical identity group. For each prime p that divides $|Q \cap S_X|$ there is a cyclic subgroup $Z_p \subset Q \cap S_X$, so there is a periodicity p in S_X caused entirely by $Q \cap S_X$.

We now consider the racemate group R_X. We know that $S_X \subset R_X$ and that S_X is a normal subgroup of R_X of index 2. Clearly, $A \cap R_X \subset R_X$, and for reasons analogous to those when we considered S_X, we also have $R_X \subset (A \cap R_X) \cdot Q$. So, using Dedekind's rule once again we find

2.2 <u>Theorem</u> Let R_X be the racemate group of a molecule X having the pattern $\{\Omega_1, \ldots, \Omega_n\}$. Then R_X splits as a semidirect product

$$R_X = (A \cap R_X) \wedge (Q \cap R_X).$$

If $A \cap R_X = A \cap S_X$, then $Q \cap R_X$ must be larger than $Q \cap S_X$.

We next introduce the notion of chemical constitution group for a molecule, a concept that does not depend on the chemical identity group of X. The chemical constitution group of X is

K_X = The set of all ligand permutations that preserve the chemical constitution of X.

Clearly, K_X is a group. Every permutation in A will preserve the connectivity list of the molecule, and therefore also its chemical constitution.

However, there may be permutations which preserve the chemical constitution of X and also interchange fans:

for example in the fan interchange (13)(24) would preserve the chemical constitution. Now, $A \subset K_X \subset A \cdot Q$, so by Dedekind's rule

2.3 Theorem Let K_X be the chemical constitution group of a flexible molecule with the pattern $\{\Omega_1, \ldots, \Omega_n\}$. Then $K_X = A \wedge (Q \cap K_X)$ is a semidirect product.

With this we can determine the number of stereoisomeric permutation isomers of X. This is

$$a = \frac{|K_X|}{|S_X|}$$

so we find

$$a = \frac{|A| \cdot |Q \cap K_X|}{|A \cap S_X| \cdot |Q \cap S_X|}$$

3. Conformationally flexible Polycentric Molecules

Polycentric molecules[*] are the subject of a major field of stereochemistry. They pose a variety of problems that have not yet been solved satisfactorily. The attempts to interpret the behaviour of such molecules

---

[*] A molecule with a polyatomic skeleton is called a polycentric molecule. Each skeletal atom is seen as a monocentric subunit of the skeleton.

in terms of geometry and point-group symmetry have not been very successful
(see the second half of this section). However, we shall show in this
section that the theory of chemical identity groups can be applied to
develop the subject on a rigorous basis and to yield results that are clear
and unambiguous.

A polycentric molecule may occur in a variety of geometric arrangements, its conformations, which are interconverted by intramolecular
motions based only on internal rotations about bond axes, which can change
the dihedral angles between the monocentric subunits of the molecule, but
which keep bond lengths practically unchanged [4]. We can therefore
consider a conformationally flexible molecule to be characterized by a
flexible polycentric molecular skeleton whose monocentric subunits are
rigid; the intramolecular motions are the internal rotations of monocentric
units around the bond axes. To have created awareness of the importance of
conformations is one of the greatest contributions of Sir D. H. R. Barton
to organic chemistry [4].

The concepts of conformation and conformer are essential to our treatment of conformational flexibility. Any snapshot of a conformationally
flexible molecule is called a conformation of that molecule. Any two conformations of a given molecule are called conformers of each other if they
are permutation isomers. It is important to note that this terminology
differs from the usual parlance. In more detail: We consider as conformers
those pictures of a polycentric molecule which have the same molecular
skeleton and which differ only with respect to the skeletal placement of
the ligands. The interconversions of the conformers are therefore represented by permutations of the ligands [2,3].

An important problem in the study of conformational flexibility has been to classify and enumerate the nonrigid isomers. Historically, two approaches have been developed, one emphasizing the physicochemical properties of such molecules, and the other quantumchemical in nature.

Following the early pioneering work of Wigner, Howard and Wilson [5], using quantum mechanical techniques on the "symmetry" of nonrigid molecules, the subject made little progress for several decades until the breakthrough that came with the spectroscopy-oriented studies of Longuet-Higgins [6] and Hougen [7]. These authors represented the "symmetry" of a nonrigid molecule by a permutation group which was to account simultaneously for the point-group symmetries of each "snapshot" so well as the "feasible" intramolecular motions.

The subsequent attempt by Stone to construct and analyze the character tables of such groups [8] did not satisfy very many chemists.

Altmann's proposal [9] to write these groups as semidirect products with one subgroup representing the feasible "intramolecular" motions, was rejected by Watson's counterexamples [10]. Nor was an alternate proposal by Woodman [11] to decompose such groups into semidirect products of a "torsional subgroup" and a "frame subgroup" entirely satisfactory, one of the reasons being that none of the latter concepts was clearly defined.

Polyà's enumeration and classification of isomeric rigid molecules [12] was extended and modified by Ruch et al [13] to include also nonrigid molecules and their interconversions. For cyclohexane and related systems this problem was attacked by Leonard, Hammond and Simmons [14], so well as by Nourse [15], and investigated very thoroughly by Frei, Bauder and Günthard [16] who introduced the concept of an isometric group of nonrigid molecules.

An analysis of the reported difficulties in the above studies indicates that for a satisfactory treatment of conformational flexibility, it is necessary to make a distinction between stereoisomers in general and stereoisomers that are at the same time permutation isomers: any given molecule must be treated as a member of a definite family of permutation isomers. This is the main reason we have insisted above that conformers must be permutation isomers. Using this more precise notion, we will show that if a molecule is known to be conformationally flexible, then by using a purely algebraic procedure its chemical identity group can be decomposed as a semi-direct product of subgroups that have clear-cut meanings.

4. The Chemical Identity Group of Conformationally Flexible Molecules

Let X be a conformationally flexible molecule. Speaking in strictly geometrical terms, its motions consist of internal rotations around bond axes combined with rotations of the entire molecule. To describe these motions heuristically, it is convenient to divide the set of all skeletal atoms into a family of pairwise disjoint sets $A_1, \ldots, A_n$ (called skeletal subunits); then the molecular motions permit no interchange of individual atoms between different skeletal subunits, and the effect of any such motion can be regarded as obtained by a combination of

1. Interchanges of the atoms within each A_i (due to the internal bond-axis rotations)

and 2. Interchanges between skeletal subunits (due to rotations of the entire molecule, or parts of the molecule).

The skeletal subunits are, in general, not uniquely determined by the above two requirements. Take a hypothetical flexible molecule such as:

which can rotate about the central atom can be considered to have five skeletal subunits, each consisting of a single atom, and it can also be considered to have the three skeletal subunits {w,y},{x,v},{u}. There are conceptual advantages in choosing the individual skeletal subunits to be so large as possible.

These are all geometrically based considerations, involving motions of the skeletal atoms. Because the chemical identity group is based entirely on ligand permutations rather than on skeletal atom motions, it cannot be expected to give information about all possible motions of the skeletal atoms. However, we are going to show that it does give complete information about all those skeletal motions that involve, or are expressible by, ligand permutations; moreover, it provides an exact way to determine systems of appropriate skeletal subunits, and has a semidirect product decomposition in which each factor represents a well-defined type of skeletal motion.

Let X be a conformationally flexible molecule with chemical identity group S and model E having the pattern $\{\Omega_1,\ldots,\Omega_n\}$. Recall that $\{1,\ldots,n\}$ is the set of monocentric skeletal sites, that each $\lambda \in S$ maps each fan Ω_i either onto itself or onto some other fan Ω_j and that each $\lambda \in A \cap S$ maps each fan Ω_i onto itself.

With each $\lambda \in S$ associate a permutation λ^* of the skeletal sites $\{1,\ldots,n\}$ by the rule

$$\lambda^*(i) = j \quad \text{if} \quad \lambda(\Omega_i) = \Omega_j$$

We call the permutation λ^* the skeletal motion represented by λ and $\{\lambda^* | \lambda \in S\}$ the set of representable skeletal motions.

Observe that a given skeletal motion may not be representable, and that different ligand permutations may represent the same skeletal motion. For example, in the hypothetical molecule

where independent rotation around each bond axis takes place, the skeletal motion represented by the blade permutation $\lambda = (12)$ keeps all the sites fixed, and can be interpreted as an internal rotation around the bond axis $\overline{tu}$; the fan permutation $\lambda = (13)(24)$ which represents a motion that interchanges the skeletal atoms t and v, can be interpreted as a rotation around the bond axis $\overline{uw}$. Since there is no $\lambda \in S$ interchanging the fans at y and z, rotation about the bond axis $\overline{xw}$ is not representable. Note that both $\lambda = (12)$ and e determine the identity skeletal motion.

From now on, we consider only the representable skeletal motions. We begin by noting that $(\lambda\mu)^* = \lambda^*\mu^*$ and $e^* = id$. Therefore, by associating with each $\lambda \in S$ the permutation λ^* of the skeletal atoms $\{1,\ldots,n\}$ we get an action of S on the set $\{1,\ldots,n\}$. Of course, this action is not necessarily effective (for example, each $\lambda \in A \cap S$ induces the identity permutation of $\{1,\dots,n\}$) and not necessarily transitive (for example, there is no λ^* sending i to j if $|\Omega_i| \neq |\Omega_j|$).

We now apply 10.3 Theorem of the Appendix to find how this action behaves. First, the set $\{1,\dots,n\}$ is uniquely partioned into transitivity domains $T_1,\dots,T_k$, with S acting transitively on each T_i. Next, by choosing an $n_i \in T_i$ for each i = 1,...,k, and then a maximal subgroup $H_i \subset G$ containing the stabilizer of n_i, each transitivity domain is partioned into blocks

$$T_i = \bigcup_{j=1}^{n_i} \Delta_{ij} \qquad i=1,\dots,k$$

and for each $\lambda \in S$, the λ^* maps every Δ_{ij} either onto itself or onto some other Δ_{ip} in the same transitivity domain. Thus, the set of blocks $\{\Delta_{ij}\}$ of skeletal atoms behave as is required of the skeletal subunits in the geometric considerations stated at the beginning of this section. We therefore call these blocks the skeletal subunits. Note that these are found by a purely algebraic process applied to the chemical identity group, rather than by any geometric consideration. Moreover, just as in the intuitive geometric discussion, the skeletal subunits are not uniquely determined: they depend on the choice of the groups H_i, i = 1,...,k. The molecule X with the skeletal subunits $\{\Delta_{ij}\}$ is denoted by $X(\Delta_{ij})$.

Again by 10.3 Theorem of the Appendix, the set F of permutations that map every Δ_{ij} onto itself is a normal subgroup of S; as indicated in the geometric discussion at the beginning of this section, F is interpreted as the group of representable bond-axis rotations; we call F the flexibility group of the molecule $X(\Delta_{ij})$. The group F evidently depends on what the skeletal subunits are (i. e. on the groups H_i we have selected); and clearly $A \cap S \subset F$.

Each coset of the factor group S/F represents a unique interchange (i. e. permutation) of the skeletal subunits. Proceeding as was done for the enveloping group N in VI,1, we can select a transversal of F in S that is a group. Therefore F is a semidirect factor of S and we have

4.1 Theorem. Let S be the chemical identity group of a conformationally flexible molecule X, let $\{\Delta_{ij}\}$ be a partition of the skeleton into skeletal subunits, and let F be the flexibility group of the molecule $X(\Delta_{ij})$. Then $S = F \wedge X_S$ is a semidirect product, where $X_S \approx S/F$ is the group of skeletal subunit interchanges that occur by using the representable geometric motions of the molecule.

As the above technical discussion indicates, we point out explicitly that the flexibility group F is not determined by the molecule alone; it also requires selecting a decomposition of the molecular skeleton into skeletal subunits. Using decompositions having different skeletal subunits will, in general, give different groups F, corresponding to the representable bond-axis rotations that the chosen set of skeletal subunits permit.

Remark. The algebraic procedure indicated above can be applied to the chemical identity group of any molecule, to produce a subgroup F; however, the interpretation of F when the molecule is not conformationally flexible is a much more delicate matter.

References

[1] see also: E. Ruch, I. Ugi, Theoret. Chim. Acta 4, 287 (1966); Topics
     Stereochem. 4, 99 (1969); J. G. Nourse, J. Amer. Chem. Soc. 101,
     1210 (1979).

[2] J. Dugundji, J. Showell, R. Kopp, D. Marquarding and I. Ugi, Isr. J.
     Chem. 20, 20 (1980).

[3] J. Gasteiger, P. Gillespie, D. Marquarding and I. Ugi, Top. Curr.
     Chem. 48, 1 (1974).

[4] D. H. R. Barton, Experientia 6, 316 (1950); Nobel Lecture: Angew.
     Chem. 82, 827 (1970); see also: M. Hanack, "Conformational Theory",
     Academic Press, New York 1965; "Conformational Analysis", G. Chiur-
     doglu ed., Academic Press, New York 1971; J. Dale, "Stereochemie und
     Konformationsanalyse", Verlag Chemie, Weinheim 1979.

[5] E. P. Wigner, Nachr. Ges. Wiss. Göttingen 1930, 130; J. B. Howard,
     J. Chem. Phys. 5, 442 (1937); E. B. Wilson, ibid. 6, 740 (1938).

[6] H. C. Longuet-Higgins, Molec. Phys. 6, 445 (1963).

[7] J. T. Hougen, J. Chem. Phys. 39, 358 (1963); Can. J. Phys. 42, 1920
     (1965); 44, 1169 (1966).

[8] A. J. Stone, J. Chem. Phys. 41, 1568 (1964).

[9] S. L. Altmann, Proc. Roy. Soc. A298, 184 (1967).

[10] J. K. G. Watson, Mol. Phys. 21, 577 (1971).

[11] C. M. Woodman, Molec. Physics 19, 753 (1970).

[12] G. Polyà, Compt. rend. Acad. Sci. Paris 201, 1176 (1935); Compt.
     rend. Acad. Sci. Paris 202, 1554 (1936); Vierteljschr. Naturforsch.
     Ges. Zürich 81, 243 (1936); Z. Krystallogr. (A) 93, 414 (1936); Acta
     Math. 68, 145 (1937); see also: N. G. DeBruijn, Koninkl. Ned. Akad.
     Wetenschap. Proc. Ser. A 62, 59 (1959); Nieuw Arch. Wiskunde (3) 18,
     61 (1970).

[13] E. Ruch, W. Hässelbarth und B. Richter, Theoret. Chim. Acta (Berl.)
     19, 288 (1970); W. Hässelbarth and E. Ruch, ibid. 29, 259 (1973);
     see also: Appendix and G. Frobenius, J. für Mathematik 4, 273 (1886);
     W. Burnside, "Theory of Groups of finite Order", Cambridge University
     Press, Cambridge 1911; J. H. Redfield, Amer. J. Math. 1927, 49;
     L. Comtet, "Analyse Combinatoire" Vol. 2, Presses Universitaires de
     France, Paris, 1970, p. 90.

[14] J. E. Leonard, G. S. Hammond and H. E. Simmons, J. Amer. Chem. Soc.
     97, 5052 (1975).

[15] J. G. Nourse, J. Chem. Inf. Comput. Sci. 21, 168 (1981).

[16] H. Frei, A. Bauder and H. H. Günthard, Top. Curr. Chem. 81, 1 (1979).

P A R T   III

APPLICATION OF THE THEORY OF THE CHEMICAL IDENTITY GROUP

TO ACTUAL CURRENT STEREOCHEMICAL PROBLEMS.

EXAMPLES, ILLUSTRATIONS AND APPLICATIONS

In this Chapter we give some applications of the theory to actual stereochemical problems. We do not intend to present a comprehensive survey; our purpose is simply to show

(a) how actual chemical problems are expressed in terms of the theory,

(b) how the theory is used to treat those problems and

(c) that the needed computations are straightforward.

Some of the given examples also indicate that the present theory has predictive power and provides criteria for judging whether or not certain experimental results are complete.

1. *The Chemical Identity Group of a Molecule with a rigid Skeleton*

1.1 Example A pentacoordinate arsenic compound with five different ligands can be represented by a model

E

It is known that under the usual observation conditions the bond system of the central atom is fairly rigid, and that the molecule behaves as though its skeleton has the D_{3h} symmetry of a trigonal bipyramid; the compound X (= 1a) represented by the model E therefore has the chemical identity group (see II.2 and IV.3).

$$S_X = \{e,(123),(132),(12)(45),(13)(45),(23)(45)\}$$

109

We will now illustrate our methods by determining several properties of this family of permutation isomers of X (=1a).

(a) According to IV,2.5, the models λE for λ ∈ S_X are all chemically identical to E, and are the only models chemically identical to E that can be obtained from E by permuting its ligands; they all represent the same permutation isomer X as does E. These are the models

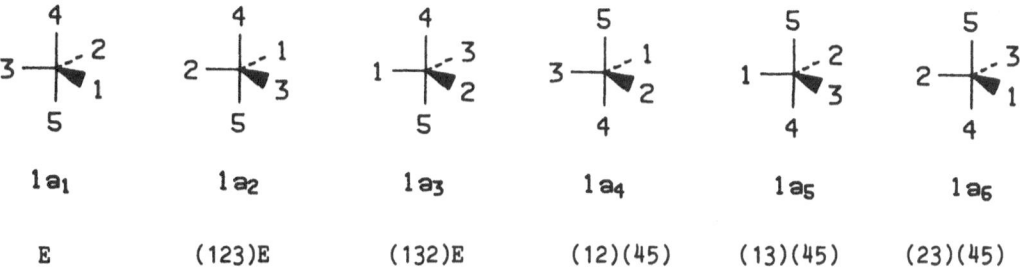

$1a_1$	$1a_2$	$1a_3$	$1a_4$	$1a_5$	$1a_6$
E	(123)E	(132)E	(12)(45)	(13)(45)	(23)(45)

and in fact, we can see directly that each is simply some rotated form of E.

(b) As stated by IV 2.5, for any permutation λ not in S_X, the model λE represents an isomer $λS_X$ that is chemically distinct from X; moreover, all the permutations in the coset $λS_X$ when applied to E will give all the models that are chemically identical to λE: they all represent to isomer λX.

Let us take λ = (14) as an example. The isomer (14)X is represented by the coset

$$(14)S_X = \{(14),(1423)(1432)(1542),(1543)(154)(23)\}$$

When these ligand permutations are applied to E, they give the models $\overline{1f_1} - \overline{1f_6}$

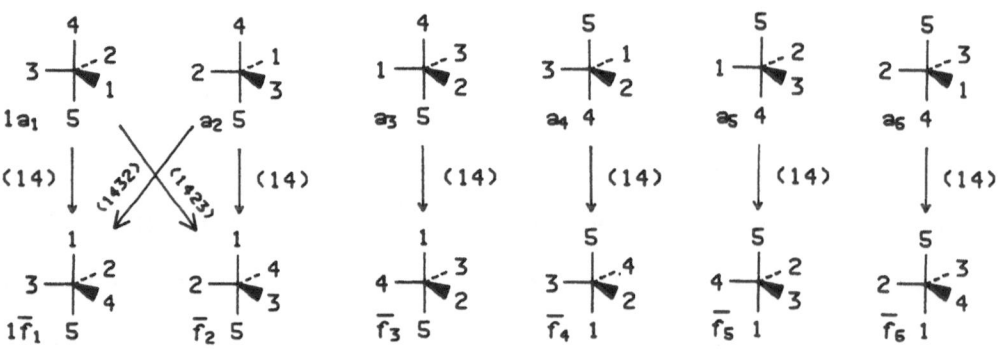

$1\bar{f}_1$ $1\bar{f}_2$ $1\bar{f}_3$ $1\bar{f}_4$ $1\bar{f}_5$ $1\bar{f}_6$

(14)E   (1423)E

which are all chemically identical, but not to $\{a_1,..,a_6\}$.

We have obtained the models $\overline{1f}_1 - \overline{1f}_6$ by first finding the coset λS_X and applying each of those permutations to E: we can equally well have applied the permutation $\lambda = (14)$ directly to each one of the models $\{a_1,..,a_6\}$ of X to get the models making up the isomer λX. In fact, we can apply any permutation in λS_X to each of the models $\{a_1,..,a_6\}$ to produce the models $\{\overline{1f}_1,..,\overline{1f}_6\}$.

(c) According to IV,3.2 the chemical identity group of the isomer (14)X will be $(14)S_X(14)^{-1}$, which is obtained by multiplying the elements of the coset $(14)S_X$ on the right by $(14)^{-1}$ or, equivalently, by conjugating S_X with (14). The latter method is simpler, since conjugates are easy to calculate (Appendix,7): in the present cases we replace in S_X each occurrence of 1 by 4 and each occurrence of 4 by 1, to get that

$$(14)S_X(14)^{-1} = \{e,(423),(432),(42)(15),(43)(15),(23)(15)\}$$

as the chemical identity group of the isomer $\overline{1f}_1$.

(d) Applying IV,2.5, the number of chemically distinct isomers in this family of permutation isomers is $[S_5:S_X] = {}^{5!}/_6 = 20$, one isomer corresponding to each coset. We list all the cosets, and a model that represents each coset:

Table 2: Family of permutation isomers 1 represented by the left cosets of S_{1a} in SymL [1-3].

1	Formula	Corresponding left coset of S_{1a} in SymL
a		e,(123),(132),(12)(45),(13)(45),(23)(45)
$\bar{a}$		(12),(13),(23),(45),(123)(45),(132)(45)
b		(124),(13)(24),(243),(254),(12543),(13254)
$\bar{b}$		(24),(1243),(1324),(1254),(13)(254),(2543)

cont'd Table 2

1	Formula	Corresponding left coset of S_{1a} in SymL
c		(125),(13)(25),(253),(245),(12453),(13245)
c̄		(25),(1253),(1325),(1245),(13)(245),(2453)
d		(134),(234),(12)(34),(13542),(354),(12354)
d̄		(34),(1234),(1342),(12)(354),(1354),(2354)
e		(135),(235),(12)(35),(13452),(345),(12345)
ē		(35),(1235),(1352),(12)(345),(1345),(2345)
f		(142),(143),(14)(23),(154),(15423),(15432)
f̄		(14),(1423),(1432),(1542),(1543),(154)(23)

cont'd Table 2

1	Formula	Corresponding left coset of S_{1a} in SymL
g	$3 \begin{smallmatrix} 1 \\ 2 \\ 5 \\ 4 \end{smallmatrix}$	(145),(14523),(14532),(152),(153),(15)(23)
$\bar{g}$	$3 \begin{smallmatrix} 4 \\ 2 \\ 5 \\ 1 \end{smallmatrix}$	(15),(1523),(1532),(1452),(1453),(145)(23)
h	$3 \begin{smallmatrix} 1 \\ 5 \\ 4 \\ 2 \end{smallmatrix}$	(14)(25),(14253),(14325),(15)(24),(15243),(15324)
$\bar{h}$	$3 \begin{smallmatrix} 1 \\ 4 \\ 5 \\ 2 \end{smallmatrix}$	(1425),(143)(25),(14)(253),(1524),(153)(24),(15)(243)
i	$5 \begin{smallmatrix} 1 \\ 2 \\ 4 \\ 3 \end{smallmatrix}$	(14)(35),(14235),(14352),(15342),(15)(34),(15234)
$\bar{i}$	$4 \begin{smallmatrix} 1 \\ 2 \\ 5 \\ 3 \end{smallmatrix}$	(1435),(14)(235),(142)(35),(152)(34),(1534),(15)(234)
j	$5 \begin{smallmatrix} 2 \\ 4 \\ 1 \\ 3 \end{smallmatrix}$	(24)(35),(12435),(13524),(12534),(13425),(25)(34)
$\bar{j}$	$4 \begin{smallmatrix} 2 \\ 5 \\ 1 \\ 3 \end{smallmatrix}$	(2435),(124)(35),(135)(24),(125)(34),(134)(25),(2534)

(e) The molecule 1a is chiral: the permutation ρ = (12) gives the model ρE

which can be interpreted to be an enantiomer of E. By IV,5.5 the enantiomer

of any model λE is the model $\lambda \rho E = \lambda \cdot (12) \cdot E$, which is the permutation λ

applied to the enantiomer of E. The enantiomers of the models $a_1, \ldots, a_6$ are

The enantiomer coset of λS_X is the coset $\lambda \rho S_X$, in Table 2, each coset

a,b,... is followed by its enantiomer coset $\bar{a}, \bar{b}, \ldots$

(f) The racemate group (IV,5.3) of the compound X is the group R_X =

$S_X \cup \rho S_X$. It is the Dieter group (IV,5.5) of the system $\{S_X, (12)S_X\}$ and

thus represents the chemical identity group of a racemic mixture of X   and

its enantiomer $\bar{X}$. By IV,5.6 each permutation in S_X interconverts the set

$\{a_1, \ldots, a_6\}$ of models, and also interconverts the set $\{\bar{a}_1, \ldots, \bar{a}_6\}$ of models.

Each permutation in (12)S_X converts the members of the set $\{a_1, \ldots, a_6\}$ to

those of $\{\bar{a}_1, \ldots, \bar{a}_6\}$. The left cosets of the racemate group are simply the

unions $a \cup \bar{a}$, $b \cup \bar{b}, \ldots$ of each coset with its enantiomer coset.

(g) In the compound X, let us make ligands 1 and 2 chemically indistin-

guishable, and also ligands 3, 4 chemically indistinguishable (i. e. L_1 =

L_2 and $L_3 = L_4$, but $L_1 \neq L_3$). To find the number of chemically distinct

isomers, we will use V,2.2. The stabilizer of this ligand substitution is

$$\Sigma \;=\; \{e,(12),(34),(12)(34)\}$$

All the permutations μ belonging to a single double coset, and only those permutations, will give models μE that belong to the same chemical compound when the ligands are made equivalent according to Σ. In particular, for any model λE, all the models μE with $\mu \in \Sigma \lambda S_\chi$ will become chemically equivalent. This double coset is easily calculated: multiply Σ by λ and take the union of all the cosets μS_χ containing those elements. Thus, for example, if $\lambda = (14)$, then $\Sigma \cdot (14) = \{(14),(12)(14),(34)(14),(12)(34)(14)\} = \{(14),(124),(143),(1243)\}$ and the cosets in Table 2 that contain these elements make up a single double coset.

The number of distinct double cosets, and the λS_χ cosets with which they are constructed is given explicitly in Table 3

116

Table 3: Family of permutation isomers 2 represented by the double cosets $\Sigma\lambda S$

Representative right Σ-coset	Member of family 1	converted into member of family 2	
e,(12)(34),(12),(34)	a,d,ā,d̄	a	(diagram: 3 — 3 / 1 / 1 / 5)
(124),(143),(14),(1243)	b,f,f̄,b̄	b	(diagram: 1 — 3 / 1 / 3 / 5)
(125),(15)(34),(15),(125)(34)	c,i,ḡ,j̄	c	(diagram: 3 — 3 / 1 / 5 / 1)
(135),(12435),(1245),(1435)	e,ē	d	(diagram: 3 — 1 / 1 / 5 / 3)
(145),(12435),(1245),(1435)	g,j,c̄,ī	c̄	(diagram: 1 — 3 / 1 / 5 / 3)
(14)(25),(15243),(1524),(143)(25)	h,h̄	e	(diagram: 1 — 3 / 5 / 3 / 1)

(h) Since there are 6 double cosets, there are exactly 6 distinct chemical compounds that can be made from X when each of two pairs of ligands are
made chemically identical. We now determine which of those compounds will
be chiral. According to V,2.3 any double coset that contains a left coset
λS_X and the enantiomeric $\lambda\rho S_X$, will be made up exclusively by such pairs,
and represents an achiral isomer; all the remaining isomers will be chiral.
As the table shows, in the case we are considering, there are exactly two
distinct chiral isomers possible. The remarks following V,2.4 give a more
direct way to determine the achirality of the compound represented by a
given double coset: the double coset $\Sigma\lambda S_X$ will represent an achiral compound if and only if the conjugate $\lambda\Sigma\lambda^{-1}$ contains the enantiomerization ρ.
Thus, in our case, the double coset $\Sigma(35)S_X$ will represent an achiral compound, since $(35)\Sigma(35)^{-1} = \{e,(12),(54),(12)(54)\}$ contains the enantiomerization $\rho = (12)$; the double coset $\Sigma(14)S_X$ will represent a chiral compound.

1.2 Example. The allene derivatives 2a - $\overline{2c}$ have in common the same
chemical identity group S_{2a} because $S_4 = \text{SymL}$ is the normalizer of S_{2a} [4].
When 2a is used as the reference isomer its permutation isomers are represented by the left cosets of S_{2a} as is indicated below. Note that S_{2a} is
also the chemical identity group of 3a.

$S_{2a,3a} = \{e,(12)(34),(13)(24),(14)(23)\}$

$(12)S_{2a,3a} = \{(12),(34),(1423),(1324)\}$

2a 3b

$(13)S_{2a,3a} = \{(13),(1432),(24),(1234)\}$

2b 3c

$(132)S_{2a,3a} = \{(132),(143),(234),(124)\}$

2b 3d

$(14)S_{2a,3a} = \{(14),(1342),(1243),(23)\}$

2c 3e

$(123)S_{2a,3a} = \{(123),(243),(142),(134)\}$

2c 3f

The racemate groups of these permutation isomers differ however:

$$R_{2a} = \overline{R_{2a}} = S_{2a} \cup \{(12),(34),(1324),(1423)\}$$

$$R_{2b} = \overline{R_{2b}} = S_{2a} \cup \{(14),(23),(1243),(1342)\}$$

$$R_{2c} = \overline{R_{2c}} = S_{2a} \cup \{(13),(24),(1432),(1234)\}$$

because the permutations which interconvert 2a - $\overline{2c}$ do not belong to the normalizer of the groups R_{2a}, R_{2b} and R_{2c} in $S_4 = SymL$ (see IV,5).

1.3 Example. Chemically distinct permutation isomers may have the same chemical identity group, but different racemate groups; and they may have the same racemate group but different chemical identity groups. There are also non-enantiomeric permutation isomers that have the same chemical identity group and the same racemate group; these are called hyperchiral isomers [4] (IV,5.9). Whether or not hyperchirality has experimentally observable consequences has been the subject of a recent debate [5] and, at the time of this writing, there is insufficient evidence to decide this matter [6].

We will show that all the above possibilities occur in the family of permutation isomers of the idealized cyclobutane derivative with eight distinguishable ligands having model

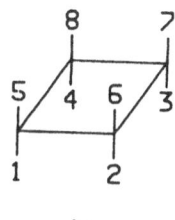

4a

This has the chemical identity group consisting of the rotations

S_X = A = {e, (13)(24)(57)(68),(15)(28)(37)(46) (16)(25)(38)(47)

        (17)(26)(35)(48),(18)(27)(36)(45),(1234)(5678),(1432)(5876)}

Thus, the family consists of $|S_8|/8$ = 8!/8 = 5040 distinct permutation isomers.

The molecule E is chiral, with ρ = (13)(57) being an enantiomerization. In Table 4, we list the enantiomer coset $\bar{A}$, and several other cosets (with their models) that will be of interest in the discussion (the bars denote enantiomers).

<u>Table 4</u>: Subfamily of permutation isomers 4 represented by the left cosets

of S_{4a}=A, the union of the normalizers of S_{4a} and R_{4a}

4	Formula	Corresponding left coset of S_{4a} in SymL
4a		A=S={e,(13)(24)(57)(68),(15)(28)(37)(46)(16) (25)(38)(47)(17)(26)(35)(48),(18)(27)(36) (45),(1234)(5678)(1432)(5876)} (rotations)
$\overline{4a}$		$\bar{A}=\bar{S}$={(13)(57),(24)(68),(12)(34)(56)(78),(14)(23) (58)(67),(15)(26)(37)(48),(17)(28)(35)(46), (1638)(2745),(1836)(2547)} (reflections and improper rotations)
4b		B={(13)(24),(57)(68),(1234)(5876),(1432)(5678), (1537)(2846),(1638)(2547),(1735)(2648), (1836)(2745)}
$\overline{4b}$		$\bar{B}$={(13)(68),(24)(57),(12)(34)(58)(67),(14) (23)(56)(78),(16)(27)(38)(45),(18)(25) (36)(47),(1537)(2648),(1735)(2846)}
4c		C={(1234),(5876),(13)(24)(5678),(57)(68) (1432),(15283746),(16253847),(17263548), (18273645)}
$\overline{4c}$		$\bar{C}$={(12)(34)(57),(13)(58)(67),(14)(23)(68), (24)(56)(78),(15483726),(16273845), (17463528),(18253647)}

Table 4 cont'd.

4	Formula	Corresponding left coset of S_{4a} in SymL
4d		D={(1432),(5678),(13)(24)(5876),(57)(68)(1234) (15463728),(16473825),(17483526),(18453627)}
$\overline{4d}$		$\bar{D}$={(12)(34)(68),(13)(56)(78),(14)(23)(57),(24) (58)(67),(15263748),(16453827),(17283546), (18473625)}
4f		F={(15)(37),(28)(46),(13)(26)(48)(57),(17)(24) (35)(68),(1256)(3478),(1458)(2763),(1674) (2385),(1872)(3654)}
$\overline{4f}$		$\bar{F}$={(17)(35),(26)(48),(13)(28)(46)(57),(15)(24) (37)(68),(1278)(3456),(1476)(2583),(1652) (3874),(1854)(2367)}
4g		G={(13)(2648),(24)(1735),(57)(2846),(68) (1537),(1258)(3476),(1456)(2783),(1654) (2387),(1852),(3674)}
$\overline{4g}$		$\bar{G}$={(13)(2846),(24)(1537),(57)(2648),(68) (1735),(1276)(3458),(1478)(2563),(1672) (3854),(1874),(2365)}

To give examples of the phenomena mentioned above, we rely on IV,5.6 – 5.8. It turns out that the normalizer of $A = S_\chi$ in S_8 is

$$N(S_\chi) = (A \cup \bar{A}) \cup (B \cup \bar{B}) \cup (C \cup \bar{C}) \cup (D \cup \bar{D})$$

and that the normalizer of the racemate group $R_\chi = (A \cup \bar{A})$ in S_8 is

$$N(R_\chi) = (A \cup \bar{A}) \cup (B \cup \bar{B}) \cup (F \cup \bar{F}) \cup (G \cup \bar{G}).$$

Thus, by IV,5.6: Because $C \in N(S_\chi) - N(R_\chi)$, the isomer 4c has the same chemical identity group as 4a, but a different racemate group.

By IV,5.7: Because $E \in N(R_\chi) - N(S_\chi)$ we find 4f and 4a have the same racemate group, but different chemical identity groups.

By IV,5.8: Because $B \in [N(R_\chi) \cap N(S_\chi)] - R_\chi$, we find 4f and 4a are non-enantiomeric, yet have the same racemate group.

2. Permutational Isomerizations of flexible Pentacoordinate Molecules

We have seen (II,5.2) that when seeking an interconversion mechanism between members of a family of permutation isomers, the Dieter group (II,5.2) determines the nature of any possible intermediary species; if the Dieter group is trivial, then there is in general no non-trivial isomerization mechanism possible.

2.1 Berry Pseudorotation

The pentacoordinate phosphorus compounds, the phosphorane derivatives 1, have a flexible skeleton. The permutational isomerizations of 1 take place by deformation of the bond angles at the central atom. In 1960 Berry

[7-9] (see also ref. [1,2]) proposed a permutation isomerization mechanism which is now called Berry pseudorotation (e. g. 1a $\leftrightharpoons$ 5 $\leftrightharpoons$ $\overline{\text{1h}}$), or just BPR [10].

The Dieter group of the system

$$D_{1a,\overline{1h}} = \text{permanent} \begin{vmatrix} S_X & (1524)S_X \\ S_X(1524)^{-1} & (1524)S_X(1524)^{-1} \end{vmatrix}$$

$$= \{e,(12)(45),(1524),(1425)\}.$$

The group $D_{1a,\overline{1h}}$ contains those permutations (like (1425) and (1524)) which interconvert 1a and $\overline{\text{1h}}$, as well as permutations (like e and (12)(45)) which preserve the chemical identities of 1a and of $\overline{\text{1h}}$.

The group $D_{1a,\overline{1h}}$ may also be interpreted as the chemical identity group of 5a, a species "half-way" between the two (no less, no more) interconverting isomers 1a and $\overline{\text{1h}}$.

If a ligand permutation λ converts a molecule X into one of its permutation isomers, and if its n-fold repetition λ^n restores the original molecule, this ligand permutation is said to have a periodicity of n [2]. The interconversion 1a $\leftrightharpoons$ $\overline{\text{1h}}$ is suitable to demonstrate that a ligand permutation may have a periodicity of two, although it is not its own inverse ($\lambda \neq \lambda^{-1}$, or $\lambda^2 \neq e$). Each one of the permutation (1425) or its inverse (1524) converts 1a into $\overline{\text{1h}}$. The order of these permutations is four, because $(1425)^4 = (1524)^4 = e$, but their periodicity is two, because

$(1425)^2 = (1524)^2 = (12)(45) \in S$, i. e. repetition of BPR (1425) or (1524) on 1a with the same pivot 3 restores 1a (rotated by 180° through $(12)(45)$).

2.2 Turnstile Rotation

A mechanistic alternative to Berry pseudorotation is "turnstile rota-tion" (TR) [1,2,9]. The isomers 1a and $\overline{1h}$ can be placed in a system larger than that consisting of the two of them alone. Assume a conversion of 1a to $\overline{1h}$ proceeds by a turnstile rotation $(153)(24)$

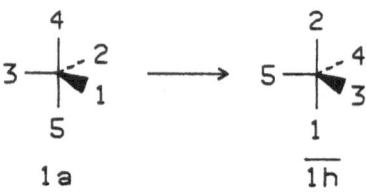

and use the isomers obtained by $[(153)(24)]^n$, n=1,...,6. This is a cyclic group C = {e,(153)(24),(135),(24)(153)(135)(24)} giving us the family

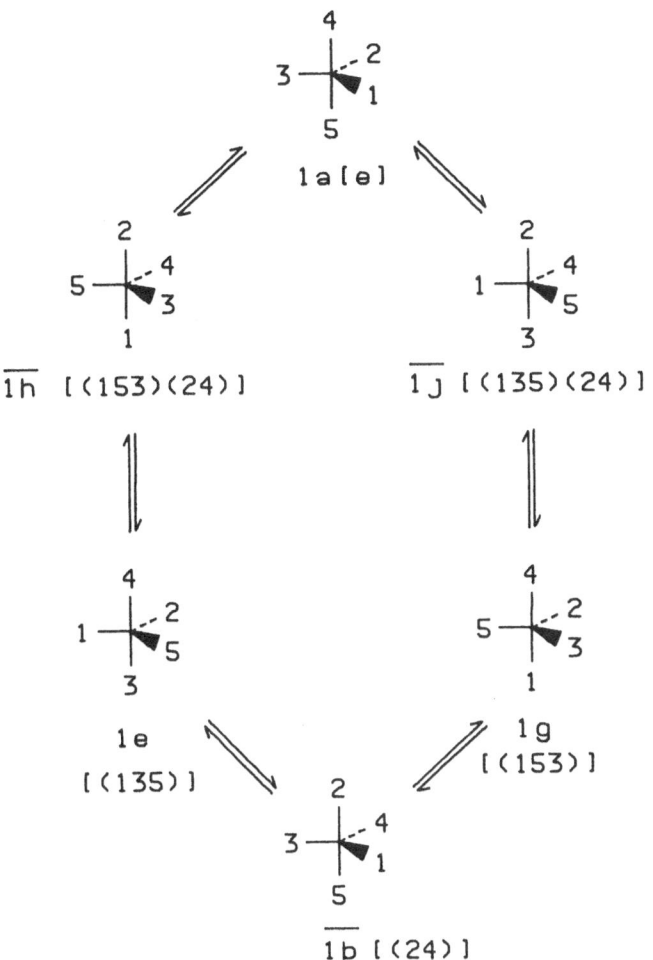

According to IV,4.5, the Dieter group of this system of isomers is non-trivial, and in fact contains the cyclic group C; thus, a nontrivial isomerization mechanism interconverting these six isomers is possible.

If there is no single molecular species that can serve in a chemically meaningful way as the intermediate in the TR-mechanism, then that inter-mediate can itself be regarded to be an ensemble of molecules with chemical identity group D[1a, $\overline{1g}$, 1g, $\overline{1b}$, 1c, $\overline{1h}$].

2.3 Double Turnstile Rotation (TR^2)

Consider the interconversion of the permutation isomers 1a, 1e and 1g. The Dieter group of this ensemble is

$$D[1a-g] = \{e,(135),(153)\}.$$

Since we have $(135) = [(153)(24)]^2$ and $(153) = [(153)(24)]^{-2}$, the interconversion of the above isomers proceeds by a process which corresponds to two successive TR whose pair and trio are the same. This process preserves the chemical identity of an ensemble of three permutation isomers and is called the double turnstile (TR^2) [1,2,9]. There exist no further ligand permutations in S_5 = SymL which also preserve the chemical identity of the considered ensemble.

2.4 The Graphs of Berry Pseudorotation and Turnstile Rotation

Berry pseudorotation of pentacoordinate phosphorane derivatives 1 proceeds via transition states of type 5 (see VII,2.1 and ref. [7-9]. Such interconversions may be traced as follows:

The permutation isomers of 1a are represented by the left cosets of S_{1a} in S_5. Thus we use the left coset space of S_{1a} as a partitioning (see Table 2). The left coset space of

$$S_{5a} = \{e,(1425),(12)(45),(1524)\}$$

serves here as a covering of S_5 (see Table 5 [3]).

Table 5: Family of permutation isomers 5 as left cosets of S_{5a} in SymL

5	Formula	left coset of S_{5a} in SymL
a	4 — 3—2 / 1 / 5	e, (1425), (1524), (12)(45)
b	4 — 2—1 / 3 / 5	(123), (14)(235), (15)(234), (23)(45)
c	4 — 1—3 / 2 / 5	(132), (24)(135), (25)(134), (13)(45)
d	3 — 4—1 / 2 / 5	(345), (14)(253), (24)(153), (12)(35)
e	4 — 5—1 / 2 / 3	(354), (15)(243), (25)(143), (12)(34)
f	4 — 3—1 / 5 / 2	(14), (125), (245), (1542)
g	2 — 3—1 / 4 / 5	(15), (254), (124), (1452)

Table 5 (cont'd.)

5	Formula	left coset of S_{5a} in SymL
h		(135),(13)(254),(1324),(13452)
i		(143),(34)(125),(2435),(15432)
j		(253),(35)(142),(1534),(12453)
k		(234),(23)(145),(1523),(12354)
l		(235),(23)(154),(1423),(12345)
m		(134),(13)(245),(1325),(13542)
n		(243),(34)(152),(1435),(12543)
o		(153),(35)(124),(2534),(14532)

Table 5 (cont'd.)

5	Formula	left coset of S_{5a} in SymL
$\bar{a}$		(12),(45),(14)(25),(15)(24)
$\bar{b}$		(23),(14235),(15234),(123)(45)
$\bar{c}$		(13),(13425),(13524),(132)(45)
$\bar{d}$		(35),(14253),(15324),(12)(345)
$\bar{e}$		(34),(14325),(15243),(12)(354)
$\bar{f}$		(25),(142),(154),(1245)
$\bar{g}$		(24),(145),(152),(1254)

Table 5 (cont'd.)

5	Formula	left coset of S_{5a} in SymL
h̄	(structure: 5—, 4, 3, 2, 1)	(13)(24),(1345),(1352),(13254)
ī	(structure: 4—, 1, 3, 5, 2)	(25)(34),(1432),(1543),(12435)
j̄	(structure: 5—, 4, 1, 3, 2)	(14)(35),(1253),(2453),(15342)
k̄	(structure: 2—, 3, 1, 4, 5)	(15)(23),(2354),(1234),(14523)
ī	(structure: 2—, 4, 1, 5, 3)	(14)(23),(2345),(1235),(15423)
m̄	(structure: 1—, 3, 4, 2, 5)	(13)(25),(1354),(1342),(13245)
n̄	(structure: 4—, 2, 1, 3, 5)	(15)(34),(1243),(2543),(14352)
ō	(structure: 1—, 4, 5, 2, 3)	(24)(35),(1532),(1453),(12534)

Table 6: The non-empty intersections of the left cosets of S_{1a} as a partition of S_5 and the left coset of S_{5a} as a covering of S_5 [3].

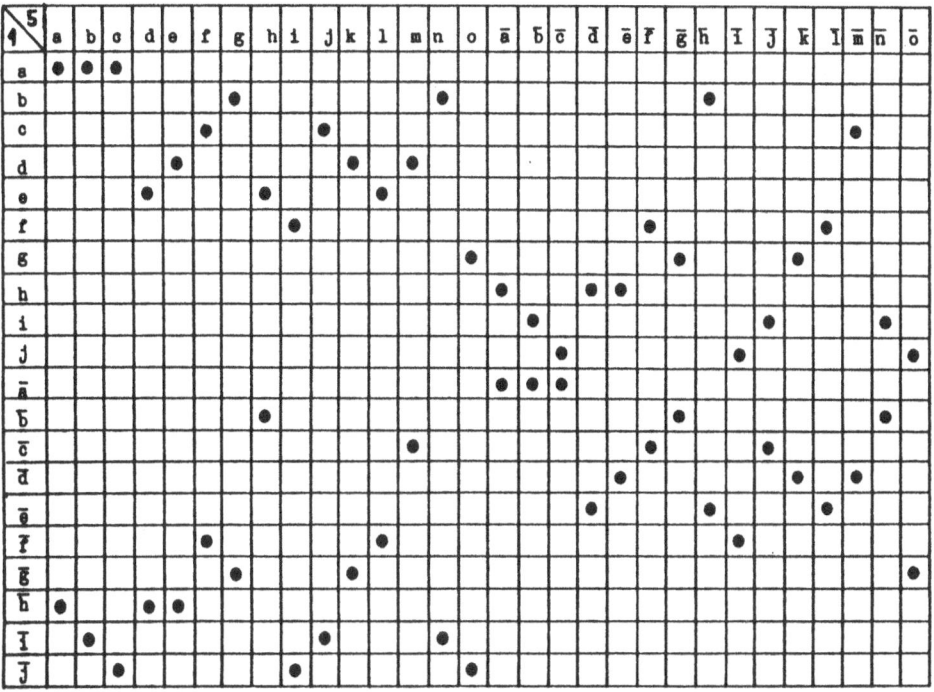

The content of Table 6 translates into Graph 1, which contains Graph 2 as a subgraph. The rows of Table 6 represent the permutation isomers of family 1, while the columns stand for the permutation isomers of family 5. A dot in row α and column β indicates that 1α and 5β are directly interconverted. This is represented by a connection between the nodes representing 1α and 5β in graph.

Graph 1: Berry pseudorotation of family 1 via family 5 (The symbols a - j̄
denote isomers of family 1 while a - ō stand for members of family 5).

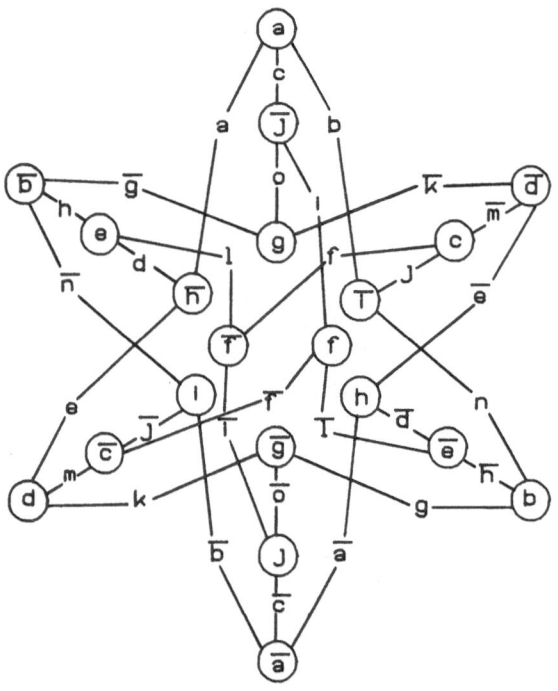

Graph 2: Berry pseudorotation of family 1 [8]

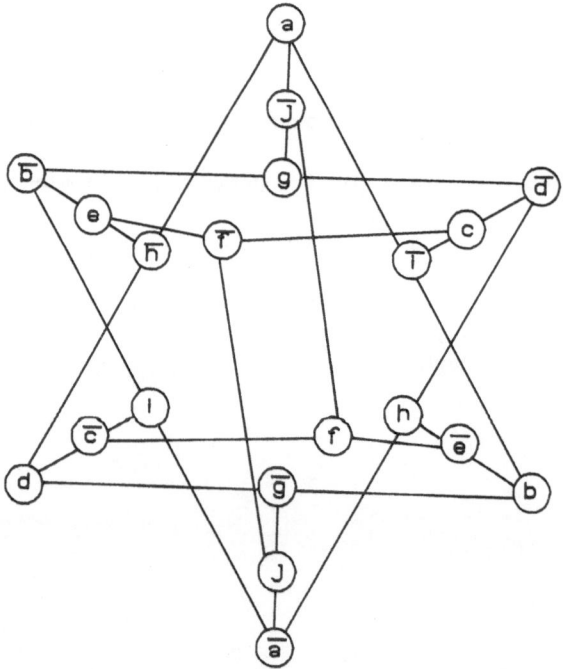

133

Graph 2 is the customary graph of the Berry pseudorotation as well as the turnstile rotation[*]), while Graph 1 contains, in addition, the respective transition states 5 between the interconverting phosphorane derivates.

Graph 2 [1,2,8] is obtained directly in an analogous manner by using the the left coset space of S_{1a} as the partition, and the left coset space of

$$S_{\overline{1h}} = \{e,(345),(354),(12)(34),(12)(35),(12)(45)\}$$

as the covering, of S_5 (see also VII,6).

Graph 3: The interconversions of the members of family 2 (Table 2) by BPR and TR are re presented by Graph 3.

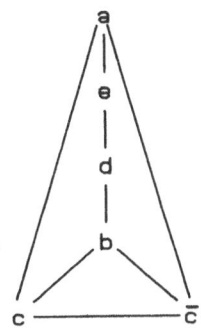

This graph is either obtained from the primary Graph 2 by removal of the nodes that are redundant, due to isomer "mergers" or Graph 3 is also obtained by using the double coset space $\Sigma\lambda S_{1a}$ as a partition of S_5, and the double coset space $\Sigma\lambda S_{\overline{1h}}$ as the covering of S_5. The intersection of the partition and the covering indicates which permutation isomers of 2a are directly interconvertible by Berry pseudorotation or turnstile rotation and correspond to the vertices of Graph 3 [1,2,12].

---

[*]) Berry pseudorotation and turnstile rotation follow the same isomerization graph, because they both belong to the same Musher mode (see V,4 and [2]).

3. Sigmatropic 1.5-Hydrogen Shift

Some years ago Roth et al. [13] designed a sophisticated experiment by which they obtained experimental evidence for the suprafacial nature of sigmatropic 1.5-shifts, showing that such processes proceed in accordance with the Woodward-Hoffmann rules [14].

They subjected the deuterated butadiene derivative 7c to the 1.5-hydrogen shift reaction and determined the essential stereochemical features of the observed products 7a and 7b. These were found to be in agreement with the predicted stereochemical course of the reaction.

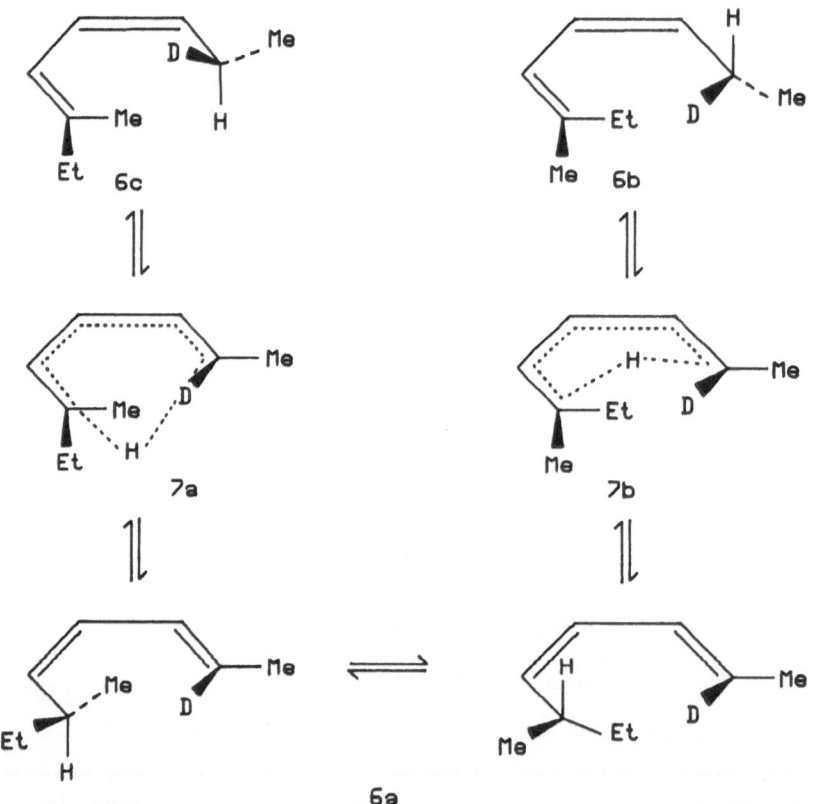

The treatment of this case by our formalism reveals that the straightforward interpretation of this experiment by Roth et al. [13] is correct,
but that it is incomplete; the situation is more complicated than origi-
nally expected. In fact, according to our results, 6a-6c are not the only
isomers which participate in the above equilibrium system: we find that the
isomer 6d must also be involved.

6d

One of the unidentified by-products of 6a - 6c may indeed be 6d. To
investigate this case by our approach, we represent the above reaction by
the following scheme:

8a (=6a)   8c (=6c)   8b (=6b)

The Dieter group of the ensemble {8a,8b,8c} is $D_{8a,8b,8c}$ = e;
According to our discussion (II,5) no nontrivial isomerization mechanism
involving only these three isomers is possible. Since the permutations
(1423) and (12)(34) should be members of $D_{8a,8b,8c}$, the ensemble of permutations isomers is too small. The group generated by the above permutations is D_8 = {e,(1324),(1423),(12)(34)}. When applied to 7a the elements
of D_7 generate the three isomers 8b, 8c and 8d. Thus the ensemble must not
only contain 8a, 8b and 8c but also 8d $\equiv$ 6d.

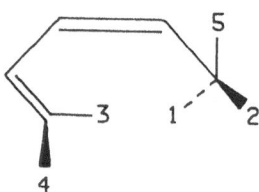

8d

Therefore, 8a, 8b, 8c and 8d interconvert by suprafacial sigmatropic 1.5-hydrogen shift. This result which we find for five chemically distinguishable ligands 1,...,5 is also valid for the case studied by Roth et al. [12] with 2 = 3 = Me, because none of the involved isomers 8a,...,8d becomes identical with another one as a consequence of the existent ligand equivalency (see VI,2).

Thus, a fourth isomer, 6d, will probably be found, when the suprafacial sigmatropic 1.5-hydrogen shifts of 6c are reexamined, and one looks for isomers beyond those already found.

4. Conformationally flexible Molecules

4.1. Example: We have seen (VI,4.7) that the chemical identity group of an acyclic conformationally flexible molecule can be expressed in terms of subgroups having clear-cut chemical meaning: one subgroup represents the internal rotations about bond axes [15], the other representing rotations of the entire molecule. We illustrate this decomposition with an ethane derivative

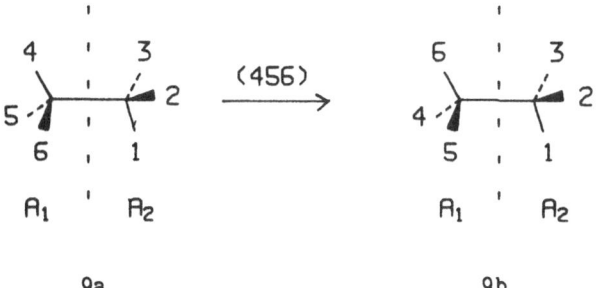

9a   9b

having the chemical identity group

S_9 = {e,(123),(132),(456),(465),(14)(26)(35),(15)(24)(36),

(16)(25)(34),(123)(456),(123)(465),(132)(456),(132)(465),

(142635),(143526),(152436),(153624),(162534),(163425)}

We take this molecule with two skeletal subunits A_1 and A_2. The flexibility

group F is the set of all permutations which map each skeletal subunit into

itself, and express the internal rotations about the C–C bond axis for any

conformation of 9. It is therefore given by blade permutations and is

F_9 = {e,(123),(132),(456),(465),(123)(456),(123)(465),(132)(456),

(132)(465)}

The factor group S/F is the group of skeletal subunit interchanges that

occur in the motions of the molecule; it is a group of order $|S|/|F|$ = 2.

According to the recipe of VI,4.1, we can choose the subgroup

$$X_{S,9} = \{e, (14)(25)(36)\}$$

and get

$$S_9 = F \wedge X_{S,9}$$

as a semidirect product decomposition.

Note that the groups S_{10}, F_{10} and $X_{S,10}$ are all identical with S_9, F_9

and $X_{S,9}$.

10

4.2 Example. The chemical identity group of the cyclohexane derivative 11 is

S_{11} = {e,(123),(132),(456),(465),(14)(26)(35),(15)(24)(36),(16)(25)(34),

(123)(456),(123)(465),(132)(456),(142635),(143526),(152436),(153624),

(162534),(163425)}

and its flexibility group is F_{11} = e, despite the fact that 11 is indeed

flexible [1,14].

11a   11b

This is the case, because the "blade permutations" at the monocentric

subunits, e. g. (17) etc., do not correspond to conformational inter-

conversions. The latter are represented by "fan permutations" such as

(165432)(7 $\widehat{12}\widehat{11}\widehat{10}$98)..*)

Under ordinary conditions a chair conformation of a cyclohexane deri-

vative like 11a or 11b is not confined to the latter conformations, but is

----

*) In order to avoid misunderstandings two-digit numbers in permutations
the constituents of two digit numbers are shown to belong together by an
arc, e. g. $\widehat{10}$.

also interconverted with conformations having other types of skeletons, e. g. the "boat", the "skew boat" etc. [15]. The chemical identity of any conformation in which 11 may exist is represented by S_{11}.

The chemical identity group of a conformationally flexible molecule contains, as subgroups, the "rigid" chemical identity groups of all conformations in which the flexible molecule can exist, as well as all ligand permutations which interconvert the conformers of a given skeletal type. From this follow all group theoretical properties of the chemical identity group of a conformationally flexible molecule, and one can take advantage of the above fact when constructing such a group.

4.3 Example. We give an example to illustrate how the result VI,4.1 that the chemical identity group of a conformationally flexible molecule has a certain structure, frequently enables us to construct and analyse the chemical identity group of such a molecule with only little effort. For the hypothetical molecule 12 all internal rotations about bond axes are representable by permutations of the ligands.

This molecule has three subunits: first, a tree carrying the ligands 1-8, secondly, a tree with the ligands 9 - 16, and finally one with the remaining ligands.

14 15 18 19

13 16 17 20

12 21

11 22

10 9 24 23

8 1

7 2

6 5 4 3

g h
i J
p
o
q
r
u
f
t
k
e
s
l
d n m a b
c

12

Internal rotations about the bond axes a-l are represented by ligand permutations belonging to (A ∩ S). This group G_0 is generated by the "blade permutations" (12),(34),(56),(78),(9 10),(11 12),(13 14),(15 16),(17 18), (19 20),(21 22),(23 24). The group G_0 is abelian because the permutations are all disjoint cycles and have order $|G_0| = 2^{12} = 4096$.

The intramolecular rotations about the bonds m-r are represented by a group G_1, which is generated by the permutations (14)(23),(58)(67),(9 12) (10 11),(13 16)(14 15),(17 20)(18 19),(21 24)(22 23). For the same reason as before, this group is abelian and is of the order $|G_1| = 2^6 = 64$.

Finally, we consider the intramolecular rotations about the bonds s,t,u. These correspond to the ligand permutations (18)(27)(36)(45),(9 16) (10 15)(11 14)(12 13) and (17 24)(18 23)(19 22)(20 21) which generate the abelian group G_2 with $|G_2| = 2^3 = 8$.

The flexibility group F is determined by all the internal bond axis rotations, and since the motions represented by G_0, G_1, G_2 are independent, F is the direct product of these three groups. Thus, $F = G_0 \times G_1 \times G_2$ and $|F| = 2^{12} 2^6 2^3 = 2,097,152$.

The rotations of the entire molecule correspond to the interchanges of the three skeletal subunits; this gives $|X_S| = 3! = 6$, so the chemical identity group S, being the semidirect product $F \wedge X_S$ has order 12,582,912.

5. Bullvalene

Bullvalene 13 consists of ten CH units connected by a fluxional system of covalent bonds. If one could "freeze" bullvalene, its structure would be represented by formula 13a.

13a   13b

At room temperature the interconversions of its momentary structures, e. g. 13a and 13b, by Cope rearrangements occur so rapidly that all of the CH subunits in bullvalene are observed to be equivalent on the time scale of NMR measurements [16].

The interpretation of bullvalene by the theory of chemical identity groups yields some noteworthy results.

Let 14a be a bullvalene derivative with ten distinguishable ligands. In 14a each number represents a ligand and a carbon atom to which it is attached.

14a   14b   14b'

The chemical identity group of 14a is

$$S_{14a} = \{e,(123)(456)(789),(132)(465)(798)\}.$$

Thus with ten distinguishable ligands there exist $|S_{10}|:|S_{14a}| = 10!:3 = 1209600$ permutation isomers in the family of 14.

The Cope rearrangement 14a → 14b can be expressed by the permutation $(1\widehat{1}0)(29)(38)(47)(56)$: this is easy to see if formula 14b is written in orientation 14b'.

Each member of the coset

$$(11̂0)(29)(38)(47)(56) \cdot S_{14a} = \{(11̂0)(29)(38)(47)(56),(11̂027548)(39),$$

$$(11̂037649)(28)\}$$

represents the Cope rearrangement 14a → 14b.

The elements of the Wigner subclass [2,17] (see V,4)

$$W = \{\lambda(11̂0)(29)(38)(47)(56)\lambda^{-1} \mid \lambda \in S_{14a}\}$$

$$= \{(11̂0)(29)(38)(47)(56),(19)(21̂0)(37)(46)(58),(18)(27)(31̂0)(45)(69)\}$$

belong to the "symmetry equivalent" [18] Cope rearrangements of 14a. The

union M of the cosets $\{wS_{14a} \mid w \in W\}$ corresponds to the Musher mode [19]

of this Cope rearrangement (see V,4); the individual cosets in M indicate

the permutation isomers of 14a which interconvert directly with 14a by Cope

rearrangements (Theorem V,4.3).

$$M = (11̂0)(29)(38)(47)(56)S_{14a} \cup (19)(21̂0)(37)(46)(58)S_{14a} \cup$$

$$\cup (18)(27)(31̂0)(45)(69)S_{14a}\}$$

The interconversions of the permutation isomers in family 14 are

therefore described by a graph whose nodes are all of degree three [20,21].

According to $(11̂0)(29)(38)(47)(56)S_{14a}$ the valence isomerizations of

bullvalene proceed with periodicities [2] of two and 14. This is reflected

by corresponding cycles in the isomerization graph of 14.

The repeated action of those Cope rearrangements of 14a that are

described by the permutations of order 14 correspond to $[(11̂027548)(39)]^2 =$

$(125̂81̂074)$ and $[(11̂037649)(28)]^2 = (13691̂074)$ of order seven. These per-

mutations are representatives of a combination of two Cope rearrangements

in which the second Cope rearrangement partially "undoes" the bond making/

breaking of the first Cope rearrangement [22]. The result is a new type of

rearrangement through which four bonds are made and four bonds are broken.

Thus the permutation (1̂1027548)(39) describes the conversion of 14a into 14b; in this process bonds 2-3,5-8,6-9 are broken, and bonds 2-5,3-6, 8-9 are made. Repetition of the action of (1̂1027548)(39) leads from 14b to 14c. Here bonds 3-6,4-7,9-10 are broken, and 3-4,6-9,7-10 are made.

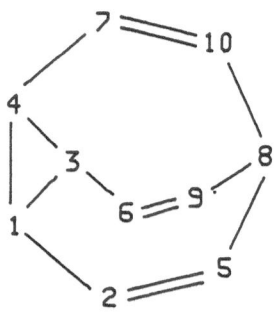

14c

The combined overall result of the aforementioned two processes is 14c. This reaction is described by [(1̂1027548)(39)]² = (12581̂074). In the process 14a → 14c the bonds 2-3,4-7,5-8,9-10 are broken, and 2-5,3-4,7-10,8-9 are made; the effect of the first Cope rearrangement on the bonds 3-6 and 6-9 is cancelled by the second Cope rearrangement. The corresponding non-zero entries in the R-matrices [23] of 15a → 15b and 15b → 15c have opposite algebraic signs [24]. The reaction 14a → 14c takes place as if a three carbon fragment 3-6-9 were rotating vs. a seven-membered tropyl-ring 15,

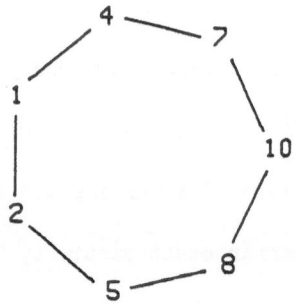

15

a process in which 3 changes its covalent connections with 1 and 2 into 1 and 4, and 9 switches bonding from 10 to 8.

The isomerization graph of 14 is obtained by using the left coset space of S_{14a} as a partition of S_{10} and the left coset space of S_{14b} as a covering, and by establishing a set valued mapping in stages:

The Cope rearrangement 14a $\rightleftharpoons$ 14b serves as a reference process. We begin the analysis with S_{14a}. In the first stage of the mapping S_{14a} intersects with $S_{14b} = \{e,(234)(567)(89\hat{1}0),(243)(576)(8\hat{1}09)\}$ $(S_{14a} \cap S_{14b} = e)$, and it hits the left coset space of S_{14b} at $(123)(456)(789)S_{14b}$ $(S_{14a} \cap (123)(456)(789)S_{14b} = (123)(456)(789) \in S_{14a})$, and at $(132)(465)(798)S_{14b}$ $(S_{14a} \cap (132)(465)(798)S_{14b} = (132)(465)(798) \in S_{14a})$. In the second stage $(123)(456)(789)S_{14b}$ hits $(13)(246)(579)(8\hat{1}0)S_{14a}$ and $(147)(\hat{1}0963)S_{14a}$ through their intersections $(13)(246)(579)(8\hat{1}0)$ and $(147)(\hat{1}0963)$; $(132)(465)(798)S_{14b}$ hits analogously $(147\hat{1}0852)S_{14a}$ and $(12)(345)(678)(9\hat{1}0)S_{14a}$. In a third stage $(13)(246)(579)(8\hat{1}0)S_{14a}$ intersects with $(258\hat{1}0963)S_{14b}$ and $(26)(459(798)S_{14b}$, $(147)(\hat{1}0963)S_{14a}$ hits $(156)(23)(489)(7\hat{1}0)S_{14b}$ and $(1562)(49)(7\hat{1}09)S_{14b}$, $(147\hat{1}0852)S_{14a}$ meets $(153)(486)(7\hat{1}09)S_{14b}$ and $(165)(23)(498)S_{14b}$, and $(12)(345)(678)(9\hat{1}0)S_{14a}$ intersects with $(135)(468)(79\hat{1}0)S_{14b}$ and $(23691\hat{0}85)S_{14b}$, etc. Conversion of the left cosets of S_{14b} into left cosets of S_{14a} by right multiplication with $(1\hat{1}0)(29)(38)(47)(56)$ would yield a graph which is labelled by left cosets of S_{14a} representing the permutation isomers of 14a.

6. *The Stereoisomers of Trihydroxyglutaric Acid*

The enumeration of the stereoisomers of trihydroxyglutaric acid 16 has been the subject of some extensive correspondence and discussion between J. van't Hoff, A. v. Baeyer and E. Fischer [26]. No general mathematical procedure has been published previously which can be used for the solution of this problem in a straightforward manner.

$$HOOC-CHOH-CHOH-CHOH-COOH$$

16

We first consider a conformationally flexible propane derivative with eight different ligands (see Table 6).

17a

Its chemical identity group is

S_{17a} = {e,(14)(26)(35)(78),(123),(162534)(78),(132),(153624)(78),(456),

(143526)(78),(465),(142635)(78),(123)(456),(16)(25)(34)(78),

(132)(465),(15)(24)(36)(78),(123)(465),(163425)(78),

(132)(456),(152436)(78)}

The group of constitution preserving ligand permutations is generated from S_{17a} and an abelian group such as

A_{17a} = {e,(12),(45),(78),(12)(45),(12)(78),(45)(78),(12)(45)(78)}

representing some of the conceivable configurational inversions. This group

$\langle A_{17a}, S_{17a} \rangle$ is of order 144. Since the order of S_{17a} is 18, we have 144/18 = 8 stereoisomers, which are represented by left cosets of S_{17a} [14].

In order to find the stereoisomers of trihydroxyglutaric acid 16 we must treat some of the ligands in 17a as equal, e. g. according to the stabilizer (see V.)

Σ = $\langle \{e,(14)\}, \{e,(257),(275),(25),(27),(57)\}, \{e,(368),(386),(36),(38),$

$\quad$ $(68)\} \rangle$.

This corresponds to the ligand equivalency 1 = 4 = COOH, 3 = 6 = 8 = OH, and 2 = 5 = 7 = H. The order of the stabilizer Σ is $2 \cdot 6 \cdot 6$ = 72.

We have the intersection $\Sigma \cap \langle A_{17a}, S_{17a} \rangle$ = $\{e,(14)(25)(36)\}$. Since these permutations belong to the left cosets of S_{17a} representing 17a and $\overline{17a}$ respectively, the latter go into the achiral trihydroxyglutaric acid 18a.

18a

$\overline{17b}$ and $\overline{17d}$ go into the chiral isomer 18b, because the left cosets of S_{17a} representing 17b and $\overline{17d}$ intersect the same right cosets of Σ, e. g. (45)(78) ϵ (45)(78)S_{17a} and (1524)(36)(78) ϵ (45)S_{17a} are elements of Σ(45)(78). The ligand permutations (45)(78) ϵ (45)(78)S_{17a} and (1524)(36)(78) ϵ (45)S_{17a} are obtained from the elements of $\{e,(14)(25)(36)\}$ = $\Sigma \cap \langle A_{17a}, S_{17a} \rangle$ by left multiplication.

18b

Similarly, one obtains $\overline{18b}$, the enantiomer of 18b, from $\overline{17b}$ and 17d. Here,

for example, the permutations $(12)e = (12) \in (12)S_{17a}$ and $(14)(25)(36)\cdot(12)$

$= (1425)(36) \in (12)(78)S_{17a}$ belong to $\Sigma(12)$.

$\overline{18b}$

The remaining stereoisomers 17c and $\overline{17c}$ go into the achiral trihydroxy-

glutaric acid 18c, since $(12)(45)e = (12)(45) \in (12)(45)S_{17}$ and

$(14)(25)(36)\cdot(12)(45) = (15)(24)(36) \in (78)S_{17a}$ are members of $\Sigma(12)(45)$.

18c

Thus, trihydroxyglutaric acid 16 exists in four stereomeric forms, one

racemate 18b and $\overline{18b}$ and two chemically achiral stereoisomers 18a and 18c.

149

Table 7: Propane derivative 17a and its stereoisomers represented by S_{17a} and its left cosets in SymL

Propane derivative 17a
and its stereo-
isomers

S_{17a} and its left cosets in SymL

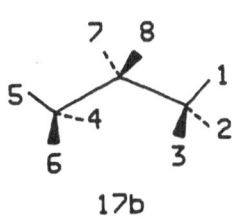

17a

e,(14)(26)(35)(78),
(123),(162534)(78),
(132),(153624)(78),
(456),(143526)(78),
(465),(142635)(78),
(123)(456),(16)(25)(34)(78),
(132)(465),(15)(24)(36)(78),
(123)(465),(163425)(78),
(132)(456),(152436)(78)

17b

(45)(78),(1435)(26),
(123)(45)(78),(1625)(34),
(132)(45)(78),(15)(2436),
(46)(78),(1426)(35),
(56)(78),(14)(2635),
(123)(46)(78),(16)(2534),
(132)(56)(78),(1524)(36),
(123)(56)(78),(1634)(25),
(132)(46)(78),(1536)(24)

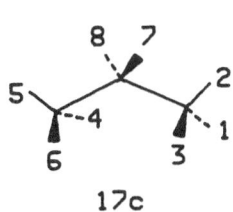

17c

(12)(45),(162435)(78),
(13)(45),(15)(26)(34)(78),
(23)(45),(143625)(78),
(12)(46),(16)(24)(35)(78),
(12)(56),(163524)(78),
(13)(46),(153426)(78),
(23)(56),(14)(25)(36)(78),
(13)(56),(142536)(78)

17d

(12)(78),(1624)(35),
(13)(78),(1534)(26),
(23)(78),(14)(2536),
(12)(456)(78),(16)(2435),
(12)(465)(78),(1635)(24),
(13)(456)(78),(1526)(34),
(23)(465)(78),(1425)(36),
(13)(465)(78),(15)2634),
(23)(456)(78),(1436)(25)

Table 7 (cont'd.)

Propane derivative 17a
and its stereo- S_{17a} and its left cosets in SymL
isomers

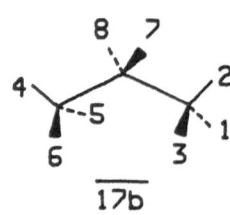

17a

(23)(56)(78),(14)(25)(36),
(12)(56)(78),(163524),
(13)(56)(78),(152634),
(23)(45)(78),(143625),
(23)(46)(78),(142536),
(12)(45)(78),(162435),
(13)(46)(78),(153426),
(12)(46)(78),(16)(24)(35),
(13)(45)(78),(15)(26)(34)

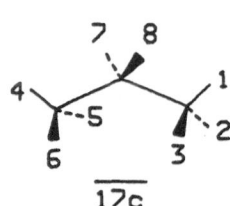

17b

(12),(1624)(35)(78),
(13),(1534)(26)(78),
(23),(14)(2536)(78),
(12)(456),(16)(2435)(78),
(12)(465),(1635)(24)(78),
(13)(456),(1526)(34)(78),
(23)(465),(1425)(36)(78),
(13)(465),(15)(2634)(78),
(23)(456),(1436)(25)(78)

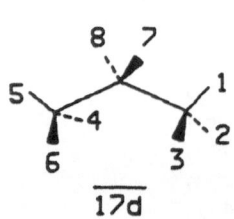

17c

(78),(14)(26)(35),
(123)(78),(162534),
(132)(78),(153624),
(456)(78),(143526),
(465)(78),(142635),
(123)(456)(78),(16)(25)(34),
(132)(465)(78),(15)(24)(36),
(123)(465)(78),(163425),
(132)(456)(78),(152436)

17d

(45),(1435)(26)(78),
(123)(45),(1625)(34)(78),
(132)(45),(15)(2436)(78),
(46),(1426)(35)(78),
(56),(14)(2635)(78),
(123)(46),(16)(2534)(78),
(132)(56),(1524)(36)(78),
(123)(56),(1634)(25)(78),
(132)(46),(1536)(24)(78)

7. S_N2 and related Processes

The theory of chemical identity groups is also applicable to molecules
that do not have a contiguous skeleton. With the present approach it is
possible to treat an ensemble of molecules as an entity with a skeleton
which consists of disjoint parts. Thus reacting chemical systems with
interconverting isomeric ensembles are also representable in terms of the
present formalism, as long as the set of ligands is the same in all partic-
ipating ensembles of molecules. We call such reactions the ligand-preserving chemical reactions.

The S_N2 [27] is a simple example of a ligand-preserving reaction.
Reaction 19a ⇌ 19b describes in general terms the substitution of a
leaving group 4 by an entering group 5 during an S_N2 type reaction [27,28].
The reaction proceeds with so-called Walden inversion [27,29] at the
asymmetric carbon atom.

19a   19b

The ligand preserving reaction between the permutationally isomeric
ensembles 19a and 19b is represented by the ligand permutation (12)(45).

The chemical identity groups of 19a and 19b are:

S_{19a} = {e,(123),(132),(124),(142),(134),(143),(234),(243),(12)(34),
     (13)(24),(14)(23)},

and the conjugate (Theorem IV,3.2)

$$S_{19b} = (12)(45)S_{19a}(12)(45) = \{e,(123),(132),(125),(152),(135),(153),$$

$$(235),(253),(12)(35),(13)(25),(15)(23)\}.$$

The Dieter group of the system $\{19a,19b\}$ is

$$D[19] = [S_{19a} \cup (12)(45)S_{19a}] \cap [S_{19b} \cup (12)(45)S_{19b}] =$$

$$= \{e,(123),(132),(12)(45),(13)(45),(23)(45)\}$$

If $D[19]$ is interpreted as the chemical identity group of an intermediate in the interconversion 19a $\rightleftharpoons$ 19b, one is led to assume the structure 1a for the intermediate "Walden species", the "watershed" of the Walden inversion.

1a

During an S_N2 type reaction 19a → 19b the entering group 5 continuously approaches the central atom while the leaving group 4 is continuously removed in a synchronized manner, until 19b is reached via 1a.

19a   20a   1a   20b   19b

Note that 20a, the intermediate of 19a $\rightleftharpoons$ 1a, and 20b, the intermediate of 1a $\rightleftharpoons$ 19b, have the same chemical identity group

$$S_{20} = \{e,(123),(132)\}$$

which is the Dieter group of $\{19a,1a\}$ and of $\{19b,1a\}$. They have also the same racemate group

$$R_{20} = S_{20} \cup \{(12),(13),(23)\}$$

and therefore (Theorem IV,5.9) 20a and 20b are hyperchiral isomers [4]; the permutations (12)(45),(13)(45),(23)(45) which formally interconvert 20a and 20b belong to the intersection $N(S_{20}) \cap N(R_{20})$ of the normalizers of S_{20} and R_{20} in S_5.

Thus the reactions 19a ⇌ 19b approach the chiral watershed 1a via the reacting species 20a and 20b which are hyperchiral isomers. When the watershed is reached 20a and 20b merge into 1a and hyperchirality vanishes. This is reminiscent of Salem's "narcissistic reactions" [30] in which a reacting system approaches an achiral watershed via enantiomeric species which merge into an achiral intermediate or transition state at the watershed.

It is interesting to note that the above formalistic representation of 19a ⇌ 20a ⇌ 1a ⇌ 20b ⇌ 19b with given starting materials and products 19a and 19b is equally valid for the reaction mechanism

$$19a \rightleftharpoons \overline{20a} \rightleftharpoons \overline{1a} \rightleftharpoons \overline{20b} \rightleftharpoons 19b$$
$$19a \rightleftharpoons 20a \rightleftharpoons \overline{1a} \rightleftharpoons 20b \rightleftharpoons 19b \text{ etc.}$$

In such a case the chemist must decide which one of the formally conceivable reaction pathways corresponds to chemical reality. In general there is always additional chemical evidence available on which a decision can be based, or there are good plausibility arguments.

We now consider 19a ⇌ 1a as a reference process for either the interconversion of the members of the family 19 through S_N2-like processes, or the permutational isomerizations by irregular processes [1] within family 1.

Table 8: The permutation isomers of 19, a tetracoordinate compound with an entering ligand, and the left cosets of S_{19a}

19   Formula   Left coset of S_{19a}

a = {e,(123),(124),(132),(134),(142),(143),
(234),(243),(12)(34),(13)(24),(14)(23)}

b = {(12)(45),(13)(45),(23)(45),(145),(245),
(345),(12345),(12453),(13245),(13452),
(14523),(14532)}

c = {(152),(153),(154),(15)(24),(15)(23),
(15)(34),(15234),(15243),(15324),(15342),
(15423),(15432)}

d = {(125),(253),(254),(13)(25),(14)(25),
(25)(34),12534),(12543),(13254),(13425),
(14235),(14325)}

e = {(135),(235),(354),(14)(35),(12)(35),
(24)(35),(12354),(12435),(13524),(13542),
(14235),(14352)}

Table 8 (cont'd.)

19   Formula   Left coset of S_{19a}

$\bar{a}$ = {(12),(13),(14),(23),(24),(34),(1234),
(1432),(1342),(1324),(1423)}

$\bar{b}$ = {(45),(123)(45),(1245),(132)(45),(1345),
(1452),(1453),(2345),(2453),(12)(345),
(13)(245),(23)(145)}

$\bar{c}$ = {(15),(1523),(1524),(1532),(1534),(1542),
(1543),(15)(234),(15)(243),(34)(152),
(24)(153),(23)(154)}

$\bar{d}$ = {(25),(1253),(1254),(1325),(25)(134),(1425)
(25)(143),(2534),(2543),(125)(34),(13)(254),
(14)(235)}

$\bar{e}$ = {(35),(1235),(124)(35),(1352),(1354),
(142)(35),(1435),(2354),(2435),(12)(354),
(24)(135),(14)(235)}

Using the left cosets of S_{21a} as a partition of S_5 and the left cosets of S_{1a} (see Table 2) as a covering of S_5, we obtain the non-empty intersection Graph 4 (see also Table 5), which allows us to trace the permutational isomerizations in families 21 and 1 by S_N2-like processes.

Graph 4:   S$_N$2-type permutational isomerizations within families 1 and 19.

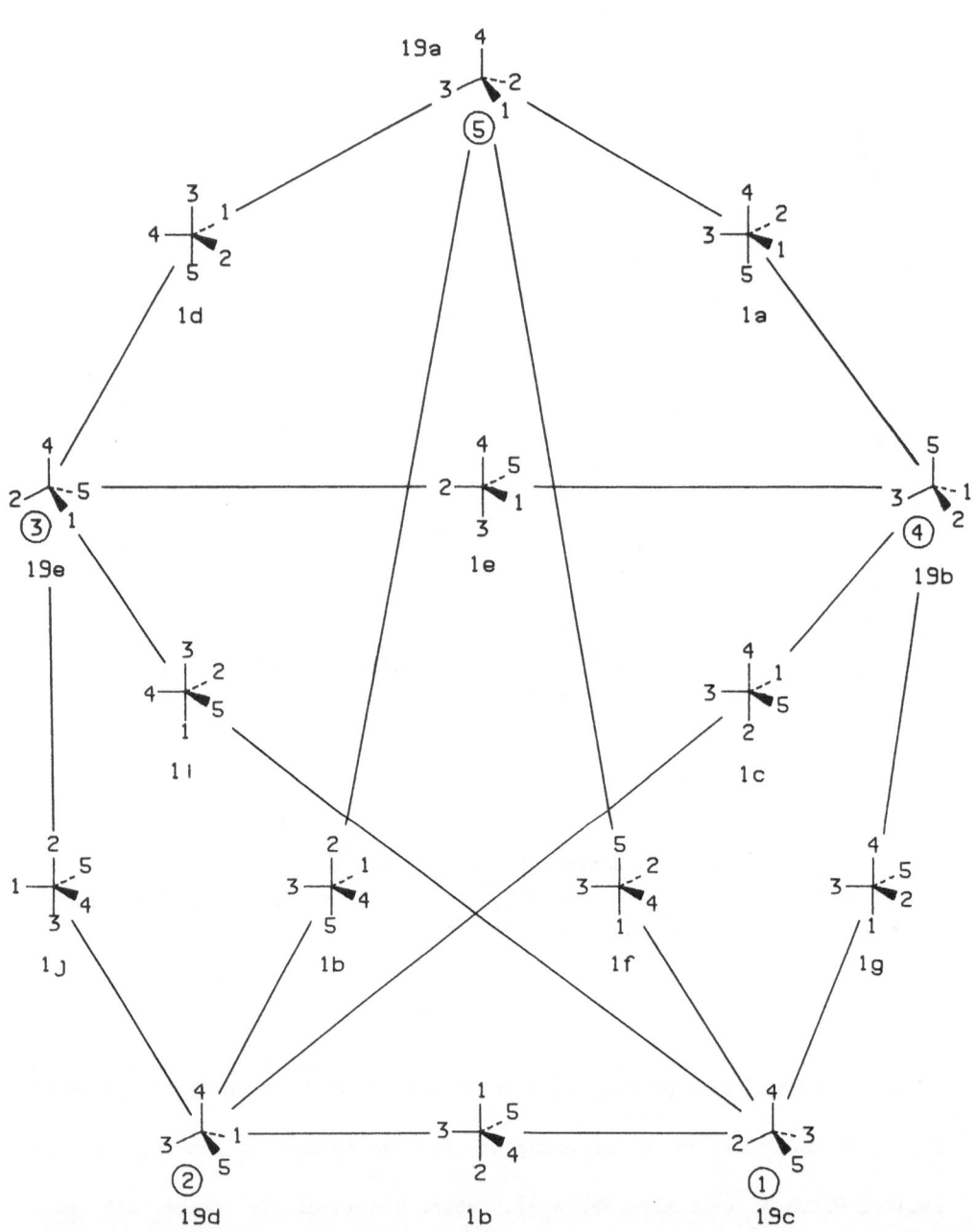

Note that this Graph 4 comprises half of family 1 and half of family 19, and that no interconversion of enantiomers is found in Graph 4. All of the permutational isomerizations in Graph 4 correspond to even ligand permutations belonging to A_5. The other halves of the families 1 and 19 are found in a disjoint "enantiomeric graph". Thus the complete permutational isomerization graph of families 1 and 19 consists of two disjoint halves, Graph 4 and its enantiomeric graph. Note that the permutational isomerizations within family 1 by $(TR)^2$ (see VII,2.3 and ref. [1,2]) is also represented by a graph consisting of two disjoint parts, without interconversion of enantiomers.

Note that somewhat different pictures result, if the Walden inversion 19a → 19b is represented by ligand permutations such as (145) and (12345) which also belong to the left coset (12)(45)S_{19a}, but are elements of other Wigner subclasses. With (145) the substitution proceeds by a mechanism characterized by the ensemble {19a,19b,19c}

having the Dieter group

$$D[19a-c] = \{e,(145),(154)\}.$$

If this Dieter group is interpreted as the chemical identity group of an intermediate, we obtain the process

Using the left coset spaces of S_{19a} and S_{1j} as a partition and a covering Graph 5 is obtained, which differs from Graph 4 by the placement of family 1.

Graph 5: Substitution processes of tetracoordinate species 19 via penta-coordinate intermediates 1.

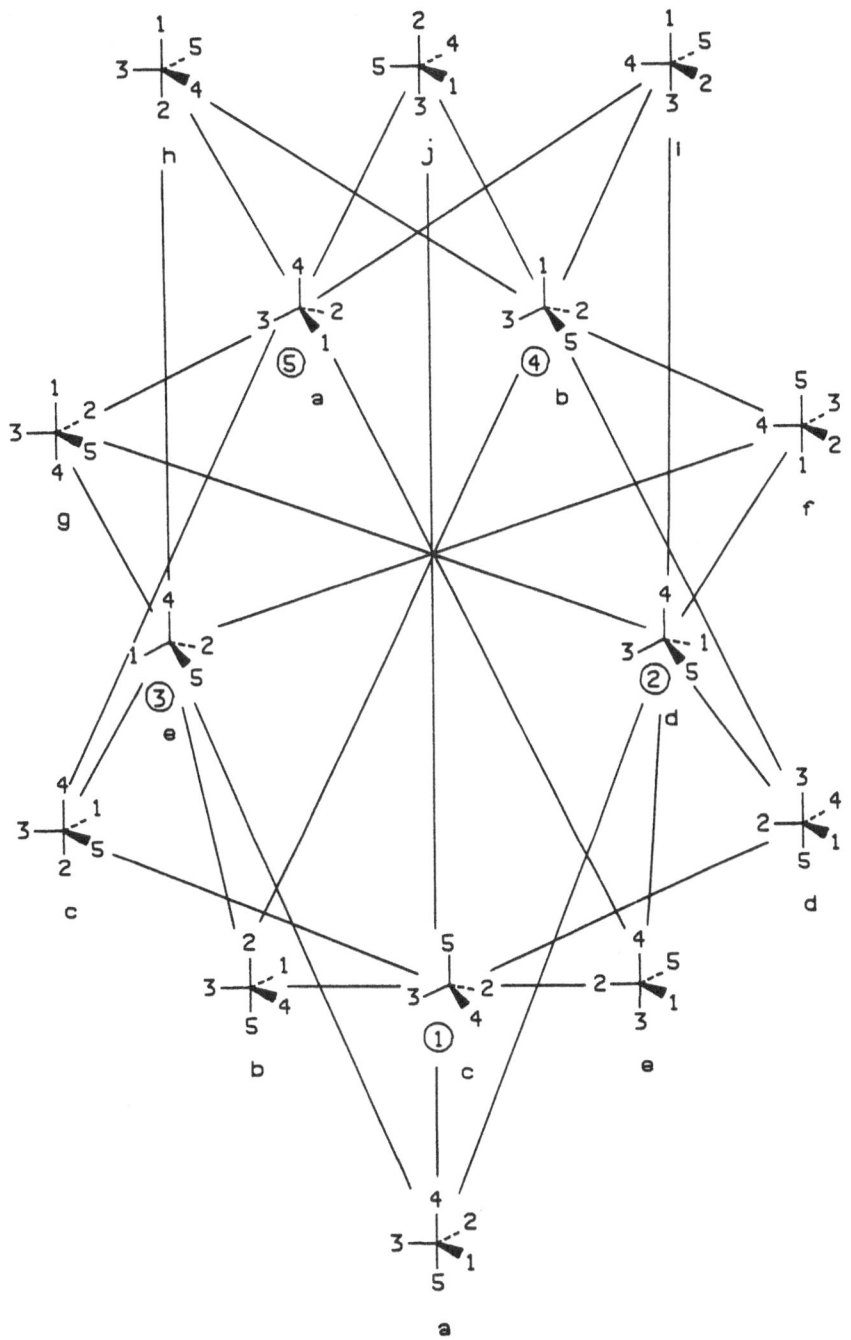

Finally, the Walden inversion 19 → 19b can be performed according to per-

mutation (12345) ∈ (12)(45)S$_{19a}$; the ensemble {19a,19b,19c,19e,19d} with

the characteristic Dieter group

$$D[19a-e] = \{(12345)^n \mid n = 1-5\}$$

results, together with still another graph representing the intercon-

version of family 19 via family 1.

If, however, the permutation isomers 19 are interconverted by sub-

stitution with "retention" [31], e.g. with 19a → $\overline{19b}$ as a reference

process,

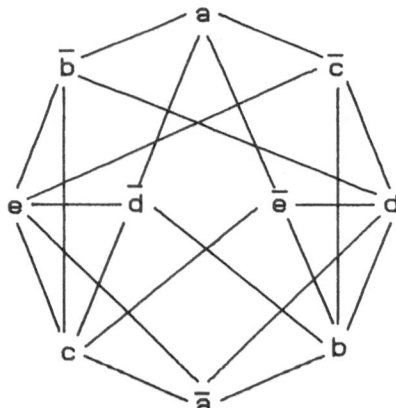

19a   19b

the interconversions within family 19 are represented by Graph 6.

Graph 6: Interconversions within family 19 by substitution with

"retention".

Note that no enantiomers are interconverted in substitution with Walden

inversion, whereas the interconversion of enantiomers takes place in sub-

stitution with "retention".

References

[1] P. Gillespie, P. Hoffmann, H. Klusacek, D. Marquarding, S. Pfohl, F. Ramirez, E. A. Tsolis and I. Ugi, Angew. Chem. $\underline{83}$, 691 (1971); Angew. Chem. Int. Ed. $\underline{10}$, 687 (1971).

[2] J. Dugundji, P. Gillespie, D. Marquarding, I. Ugi and F. Ramirez, in: "Chemical Applications of Graph Theory", ed.: A. T. Balaban, Academic Press, London 1976, p. 107.

[3] R. Kopp, Dissertation, Techn. Universität München, 1979.

[4] J. Dugundji, D. Marquarding and I. Ugi, Chemica Scripta a) $\underline{9}$, 74 (1976); b) $\underline{11}$, 17 (1977).

[5] W. Hässelbarth, Chem. Scripta $\underline{10}$, 97 (1976); $\underline{11}$, 148 (1977); C. A. Mead, ibid. $\underline{10}$, 101 (1976); $\underline{11}$, 145 (1977); see also: G. Derflinger and H. Keller, Theoret. Chim. Acta (Berl.) $\underline{56}$, 1 (1980);

[6] P. Lemmen, Chem. Ber. $\underline{115}$, 1902 (1982); Liebigs Ann. Chem. (in press); M. Noltemeyer and P. Lemmen (in preparation).

[7] R. S. Berry, J. Chem. Phys. $\underline{32}$, 933 (1960).

[8] see also: E. L. Muetterties, W. Mahler and R. Schnitzler, Inorg. Chem. $\underline{2}$, 613 (1963); E. L. Muetterties and R. A. Schunn, Quart. Rev. (London) $\underline{20}$, 245 (1966); D. Hellwinkel, Chem. Ber. $\underline{99}$, 3628, 3660 (1966); Angew. Chem. $\underline{78}$, 749 (1966); Angew. Chem Int. Ed. $\underline{5}$, 725 (1966); F. H. Westheimer Acc. Chem. Res. $\underline{1}$, 70 (1968); F. Ramirez, Acc. Chem. Res. $\underline{1}$, 168 (1968); M. J. Gallagher and I. D. Jenkins, in: "Topics in Stereochemistry", Vol. 3, ed.: N. L. Allinger and E. L. Eliel, J. Wiley & Sons, New York 1968, p.1; R. R. Holmes and R. M. Deiters, J. Amer. Chem. Soc. $\underline{90}$, 5021 (1968); Inorg. Chem. $\underline{7}$, 2229 (1968); P. C. Lauterbur and F. Ramirez, J. Amer. Chem. Soc. $\underline{90}$, 6722 (1968); J. D. Dunitz and V. Prelog, Angew. Chem. $\underline{80}$, 700 (1968); Angew. Chem. Int. Ed. $\underline{7}$, 725 (1968); E. L. Muetterties, J. Amer. Chem. Soc. $\underline{91}$, 1636, 4115 (1969); G. M. Whitesides and H. L. Mitchell, J. Amer. Chem. Soc. $\underline{91}$, 5384 (1969); K. E. De Bruijn, K. Naumann, G. Zon and K. Mislow, J. Amer. Chem. Soc. $\underline{91}$, 7031 (1969); M. Gielen and J. Nasielski, Bull. Soc. Chim. Belges $\underline{78}$, 339 (1969); M. Gielen, Medel. Vlaam. Chem. Ver. $\underline{31}$, 185, 201 (1969); K. Mislow, Acc. Chem. Res. $\underline{3}$, 321 (1970); D. Z. Denney, D. W. White and D. B. Denney, J. Amer. Chem. Soc. $\underline{93}$, 2066 (1971); J. Brocas, Theor. Chim. Acta (Berl.) $\underline{21}$, 79 (1971); J. Brocas and M. Gielen, Bull. Soc. Chim. Belges $\underline{80}$, 207

(1971); R. Hoffmann, J. M. Howell and E. L. Muetterties, J. Amer.
Chem. Soc. <u>94</u>, 3047 (1972); R. R. Holmes, Acc. Chem. Res. <u>5</u>, 296
(1972); A. Rauk, L. C. Allen and K. Mislow,
J. Amer. Chem. Soc. <u>94</u>, 3035 (1972);
J. Brocas and R. Willem, Bull. Soc. Chim. Belges <u>82</u>, 469 (1973);
L. S. Bartell and V. Plato, J. Amer. Chem. Soc. <u>95</u>, 3097 (1973);
E. L. Muetterties and L. J. Guggenberger, ibid. <u>96</u>, 1748 (1974);
M. Eisenhut, H. L. Mitchell, D. D. Traficante, J. M. Deutsch
and G. M. Whitesides, ibid. <u>96</u>, 5385 (1974); J. Demuynck, A. Strich
and A. Veillard, Nouv. J. Chim. <u>1</u>, 217 (1977); R. R. Holmes and J. A.
Deiters, J. Amer. Chem. Soc. <u>99</u>, 3318 (1977); J. A. Deiters, J. C.
Gallaci, T. E. Clark and R. R. Holmes, ibid. <u>99</u>, 5461 (1977); R. R.
Holmes, J. A. Deiters and J. C. Gallaci, ibid. <u>100</u>, 7393 (1978);
R. R. Holmes, Acc. Chem. Res. <u>12</u>, 257 (1979).

[9] M. Gielen and N. Vanlautem, Bull. Soc. Chim. Belges <u>79</u>, 679
(1970); F. Ramirez, S. Pfohl, E. A. Tsolis, J. F. Pilot, C. P. Smith,
I. Ugi, D. Marquarding, P. Gillespie and P. Hoffmann, Phosphorus <u>1</u>, 1
(1971); I. Ugi, D. Marquarding, H. Klusacek, P. Gillespie and F.
Ramirez, Acc. Chem. Res. <u>4</u>, 288 (1971); F. Ramirez and I. Ugi, in:
"Advances in Physical Organic Chemistry", ed.: V. Gold, Academic
Press, London 1971, p. 25; I. Ugi and F. Ramirez, Chem. in Britain <u>8</u>,
198 (1972); A. T. Balaban, Rev. Roum. Chim. <u>18</u>, 855 (1973); R. Lucken-
bach, "Dynamic Stereochemistry of Pentacoordinated Phosphorus and
Related Elements", G. Thieme Verlag, Stuttgart 1973; F. Ramirez,
I. Ugi, F. Lin, S. Pfohl, P. Hoffmann and D. Marquarding, Tetrahedron
<u>30</u>, 371 (1974); S. Tripett, ed.: "Organophosphorus Chemistry", Vol.
1-6, Specialist Periodical Reports, The Chemical Society, London 1969-
1975; W. E. McEwen and K. D. Berlin, eds., "Organophosporus Stereochemistry", Parts I and II, Dowden, Hutchinson & Ross, Stroudsburg,
Pa. 1975; J. Emsely and D. Hall, "The Chemistry of Phosphorus", Wiley,
New York 1976; M. Gielen, in: "Chemical Applications of Graph Theory",
ed.: A. T. Balaban, Academic Press, London 1976, p. 261; J. A. Altmann, K. Gates and J. G. Csizmadia, J. Amer. Chem. Soc. <u>98</u>, 1450
(1976); W. S. Sheldrick, Top. Curr. Chem. <u>73</u>, 1 (1978);
D. E. C. Corbridge, "Phosphorus", Elsevier, New York 1978;
D. J. H. Smith, in: "Comprehensive Organic Chemistry", Vol. 2,
eds.: D. H. R. Barton and W. D.Ollis, Pergamon, Oxford 1979,

p. 1233; S.-K. Shih, S. Peyerimhoff and R. J. Buenker,
J. Chem. Soc. Faraday Trans.II 75, 379 (1979); D. B. Denney,
D. Z. Denney, D. M. Gavrilovic, P. J. Hammond, C. Huang and K.
S.-K. Tseng, J. Amer. Chem. Soc. 102, 7072 (1980); R. R. Holmes,
"Pentacoordinated Phosphorus", Vol. I,II, ACS Monographs 175 and 176,
American Chemical Society, Washington, D.C. 1980; J. Brocas, M. Gielen
and R. Willem, "The Permutational Approach to Dynamic Stereo-
chemistry", McGraw-Hill, New York 1983 and references therein.

[10] The term "pseudorotation" has already been assigned to a cyclic conformational puckering motion of cyclopentane derivatives [11].
Accordingly, processes like 1a → 1h should be called **Berry
pseudorotation** (BPR), in order to avoid confusion in terminology.

[11] J. E. Kilpatrick, K. S. Pitzer and R. Spitzer, J. Amer. Chem. Soc. 69,
2483 (1947); see also: R. L. Hildebrandt and Q. Shen, J. Phys. Chem.
86, 587 (1982).

[12] see also: J. G. Nourse, J. Amer. Chem. Soc. 99, 2063 (1977).

[13] W. R. Roth, J. König and K. Stein, Chem. Ber. 103, 426 (1970);
see also: W. R. Roth, Chimia 20, 229 (1966).

[14] R. B. Woodward and R. Hoffmann, "Die Erhaltung der Orbitalsymmetrie",
Verlag Chemie, Weinheim 1970; see also: T. L. Gilchrist and R. C.
Storr, "Organic Reactions and Orbital Symmetry", Cambridge University
Press, Cambridge 1979, second edit.; I. Fleming, "Grenzorbitale und
Reaktionen organischer Verbindungen", Verlag Chemie, Weinheim 1979.

[15] J. Dugundji, J. Showell, R. Kopp, D. Marquarding and I. Ugi, Isr. J.
Chem. 20, 20 (1980).

[16] W. v. E. Doering and W. R. Roth, Angew. Chem. 75, 27 (1963); Angew.
Chem. Int. Ed. Engl. 2, 24 (1963); G. Schröder, ibid. 75, 722 (1963);
2, 694 (1963); J. F. M. Oth, R. Merenyi, G. Engel and G. Schroeder,
Tet. Lett. 1966, 3377; J. F. M. Oth, R. Merenyi and H. Röttele,
Chem. Ber. 100, 3538 (1967); G. Binsch, E. L. Eliel and H. Kessler,
Angew. Chem. 83, 618 (1971); Angew. Chem. Int. Ed. 10, 570 (1971); G.
Schröder, Chem. Ber., 97 3140 (1964); H. Günther, H. Klose and G.
Wendisch, Tetrahedron 25, 1531 (1969); G. Schröder, J. F. M. Oth and
R. Merenyi, Angew. Chem. 77, 774 (1965); Angew. Chem. Int. Ed. 4,
752 (1965).

[17] E. P. Wigner, "Spectroscopic and Group Theoretical Methods in Physics
(Racah Mem. Vol.), North Holland Publ. Co. Amsterdam 1971, p. 131;

Proc. Roy. Soc. (London) A322, 181 (1971).

[18] E. Ruch and W. Hässelbarth, Theoret. Chim. Acta 29, 259 (1973).

[19] J. I. Musher, J. Amer. Chem. Soc. 94, 5662 (1972); Inorg. Chem. 11,
2335 (1972); J. Chem. Educ. 51, 94 (1974); see also: M. Gielen, J.
Brocas, M. De Clerq and G. Mayence, Proc. of the 3. Symp. Coord.
Chem., Vol. 1, Ed. M. T. Beck, Brussels 1970, p. 495; M. Gielen and N.
VanLautem, Bull. Soc. Chim. Belges 79, 679 (1970); 80, 207 (1971); J.
Brocas, Top. Curr. Chem. 32, 44 (1972); J. Brocas and R. Willem, Bull.
Soc. Chim. Belges 82, 469, (1973); J. Brocas, D. Fastenakel,
J. Hicquebrand and R. Willem, ibid. 82, 629 (1973); D. J. Klein
and A. H. Cowley, J. Amer. Chem. Soc. 97, 1633 (1975);
J. G. Nourse, ibid. 99, 2063 (1977).

[20] O. Ore, "Theory of Graphs", Amer. Math. Soc., Providence, R. I. 1962;
F. Harary, "Graph Theory", Addison-Wesley, Reading 1969.

[21] A. T. Balaban, ed., "Chemical Applications of Graph Theory", Academic
Press, London 1976.

[22] L. A. Paquette, T. J. Barton and E. B. Whipple, J. Amer. Chem. Soc.
89, 5481 (1967).

[23] J. Dugundji and I. Ugi, Top. Curr. Chem. 39, 19 (1973).

[24] This fact was called to our attention by J. Brandt, who had noticed it
during his work with a novel type of hierarchic reaction documentation
program [25].

[25] J. Brandt, J. Bauer, R. M. Frank and A. von Scholley, Chem. Scripta
18, 53 (1981).

[26] E. Fischer, "Aus meinem Leben", Springer Verlag, Berlin 1922, p. 134;
see also: E. Fischer, Ber. dtsch. chem. Ges. 24, 1836 (1891);
V. Prelog and G. Helmchen, Helv. Chim. Acta 55, 2581 (1972).

[27] C. K. Ingold, "Structure and Mechanism in Organic Chemistry", Cornwall
Press, Ithaca 1953, p. 304, 509.

[28] see also: G. W. Fenton and C. K. Ingold, J. Chem. Soc. 1929, 2342; G.
F. Fenton, L. Hey and C. K. Ingold, ibid. 1933, 989; K. F. Kumli, W.
E. McEwen and C. A. Vander Werf, J. Amer. Chem. Soc. 81, 3805 (1959);
A. Bladé-Font, C. A. Vander Werf and W. E. McEwen, ibid. 82, 2396,
2646 (1960); C. B. Parisek, W. E. McEwen and C. A. Vander Werf, ibid.
82, 5503 (1960); W. E. McEwen, A. Bladé-Font and C. A. Vander Werf,
ibid. 84, 677 (1962); W. E. McEwen, K. F. Kumli, A. Bladé-Font, M.
Zanger and C. A. Vander Werf, ibid. 86, 2378 (1964); D. Marquarding,

F. Ramirez, I. Ugi and P. D. Gillespie, Angew. Chem. 85, 99 (1973); Angew. Chem. Int. Ed. 12, 91 (1973).

[29] P. Walden, Ber. dtsch. chem. Ges. 29, 133 (1896); 30, 3146 (1898); ibid. P. Walden and O. Lutz 30, 2795, (1898); P. Walden, ibid. 32, 1833, 1855 (1899).

[30] L. Salem, Acc. Chem. Res. 4, 322 (1971).

[31] see e. g.: K. E. De Bruijn, K. Naumann, G. Zon and K. Mislow, J. Amer. Chem. Soc. 91, 7031 (1969); K. Mislow, Pure and Appl. Chem. 25, 549 (1971); D. J. Cram and J. M. Cram, Top. Curr. Chem. 31, 1 (1972); E. W. Covin, A. K. Beck, B. Bastani, D. D. Seebach, Y. Kai and J. D. Dunitz, Helv. Chim. Acta. 63, 697 (1980).

C H A P T E R   VIII

A UNIFIED NOMENCLATURE AND CHEMICAL DOCUMENTATION SYSTEM

1. Desirable Features of a Chemical Documentation System

It is obvious that in order to have efficient and reliable exchange of information, a system of chemical nomenclature and documentation must accurately describe and permit exact reconstruction of the essential features of molecules (including their stereochemical features) from the encoded information.

The enormous mass of accumulated data and the ever-increasing complexity of molecular structures being studied make the use of large-scale computers mandatory for chemical documentation. In order to fully and effectively utilize the capabilities of such computers, the documentation program designed for computer processing should be efficient in terms of data entry, storage, retrieval, and adaptability.

The data entry problem is not confined merely to cost-cutting and time-saving methods, but involves the effective avoidance of ambiguities and redundancies, a minimal loss of information in compiling entry data, the ability to enter data directly with a minimal amount of human effort and ad hoc treatment of individual cases. In storage and retrieval, a chemical documentation system should accept and deliver all the data required, with as little unwanted data as possible. It should allow any type of recorded data to serve as a key for retrieval, and it should in particular be capable of retrieving the essential constitutional and stereochemical features of molecules regardless of the momentary individual geometry in which they are specified. Moreover, the system should permit the comparison of structural features of molecules and the determination of analogies. Finally, it should be adaptable to answer a large variety of questions, even those whose importance may not yet have been foreseen at the time the system was implemented. Such a computer-oriented system should also provide an interface between the documentation of the constitutional and the stereochemical aspect, and also afford an interface with a documentation of the constitutional [1] as well as the stereochemical aspects of chemical reactions. All these minimal requirements clearly indicate that a system of nomenclature and documentation should be based on a logical and consistent model of chemistry that applies to a broad range of molecules. In what follows, we intend first to examine in broad outline the basis of the traditional nomenclature systems [2-8] and point out some of their inherent weaknesses, which make it unlikely that any one of them can be readily modified to satisfy the requirements specified above. We then propose an alternative nomenclature and documentation system based on CANON [9,10] and BE-matrices [11] and on the concept of permutational isomerism [12-14]. It appears to have many - if not all - the desirable features we have detailed. Moreover, this permutational nomenclature is generally compatible with the presently used R,S-nomenclature, because the individual descriptors in both nomenclature systems will either agree in most cases, and in those cases in which they do not they can be mutually computertranslated.

2. *Some Remarks on Traditional Nomenclature and Documentation*

It is customary to confine chemical documentation and nomenclature of a molecule to the structural features that characterize its chemical identity. The commonly used features are the empirical molecular formula, constitution, and, for stereochemical purposes, configuration and conformation. These notions, although they cannot be objectively and clearly defined in all cases, still provide the most suitable conceptual framework available for chemical nomenclature and documentation because they can be used to construct a hierarchic order of equivalence classes in the set of all molecular structures [15]: any two molecules with the same molecular formula may differ in constitution (and are then called constitutional isomers); constitutional isomers may differ in configuration (and are called configurational isomers = the classical stereoisomers). Further, they may be stereoisomers differing in conformation (conformational isomers). This hierarchy is quite natural because it correlates roughly with the energy barriers to interconversion [15].

In practice, the conformational features are largely neglected in nomenclature considerations because they are generally not time-invariant: under ordinary observation conditions, conformational changes occur so rapidly that distinct conformations can only rarely be isolated or individually observed. Thus, the problem of chemical documentation reduces to that of adequately describing the constitution and configuration and of molecules that behave as if they had rigid skeletons.

Experience has shown that neither the conventional constitutional nomenclature system [16] nor any of the fragment codes [17] are very well suited for modern computer-assisted constitutional documentation systems. At present, a modified version of the Morgan algorithm [18] is used to index the atoms and thereby describe the chemical constitution in terms of canonical connectivity tables. Some previous versions of the Morgan algorithm were not free of ambiguities. We note that the published proof of uniqueness of the Morgan indices [19b] also contains ambiguities arising from the ambiguous definitions used. Moreover, the indexing is not easily adapted to further computer processing. Morgan's procedure does not take into account structural properties. Clearly, this is a serious weakness of the currently used chemical documentation systems which are based on Morgan indices, in particular with respect to stereochemistry [19,20].

The asymmetric carbon atom as a configurational unit of stereoisomers was introduced by van't Hoff [15,21], and the influence of this chirality-oriented [22] concept is still so strong that many configurational problems are treated on the basis of this model without questioning its pertinence for the given case.

In the nomenclature system currently used, the asymmetric carbon atoms are considered to be the configurational or 'stereogenic subunits' [5,23] responsible for stereoisomerism. The stereochemical features of molecules are then described in terms of the configurations of asymmetric carbon atoms or chiral configurational subunits for which an analogy with the asymmetric carbon atom is established. With n configurational units, one determines for each of these units which of the two possible configurations it has, and assigns a corresponding local descriptor. Thus, a given stereoisomer is essentially named by a descriptor consisting of n parts.

The basic justification for this stereochemical nomenclature is flawed, because we know since Aschan's critisism of van't Hoff [24] that chirality is not a prerequisite for stereoisomerism. Moreover, a nomenclature based on configurational subunits can be valid only if these subunits are independent, otherwise a variety of complications, redundancies and ambiguities can arise [15]. And indeed there exist cases where there are n configurational subunits but, instead of the 2^n expected stereoisomers there are in some instances more than and in other instances fewer than, 2^n stereoisomers [15]. Thus, the present system, although attractive, is not uniformly applicable and does not seem to be easily modifiable without a plethora of additional rules, which would make it even more cumbersome. Even then the reliability of such a system would still be questionable because one would not know, for example, how many stereoisomers a given compound has. In addition, a nomenclature system based on local configurational units does not always adequately express the fact that the constitutional and the stereochemical features of a molecule are intimately related, sometimes inseparably.

The above remarks indicate that the chemical documentation systems currently used seem to lack many of the features that are desirable in an ideal system and that the known difficulties cannot all be overcome by minor surgery [5]. We shall discuss a unified approach to constitutional so well as stereochemical nomenclature and documentation which avoids some of the flaws of the systems currently used.

3. *Representation of the Constitutional Aspect of Molecules*

The most efficient methods of computer-oriented constitutional documentation are all based on representing the chemical constitution of molecules by matrices. Adjacency, connectivity [25] and BE- (bond and electron) matrices [11,26] can be used for this purpose.

In all these matrices the row/columns are assigned to the atoms or atomic cores of the molecule to be represented. In an adjacency matrix $A = \langle a_{ij} \rangle$ an entry $a_{ij} = 1$ in the i-th row and j-th column indicates that the atoms A_i and A_j are connected by a covalent bond, and $a_{ij} = 0$ means that they are not connected. In the corresponding connectivity and BE-matrices, $C = \langle c_{ij} \rangle$ and $B = \langle b_{ij} \rangle$, the entries c_{ij} and b_{ij} are the formal order of the covalent bond between A_i and A_j; e.g. $b_{ij} = b_{ji} = 2$ would mean there is a covalent double bond between A_i and A_j.

The diagonal of an adjacency or connectivity matrix is normally a vector of atomic symbols which represents the assignment of the atoms to the rows/columns of the matrix. In a BE-matrix the diagonal entry b_{ii} is the number of free valence electrons belonging to A_i. The correlation of the atomic cores with the rows/columns of a BE-matrix is stated separately by an atomic vector.

In chemical documentation, the BE-matrices have some advantages over the other types of matrices. With its diagonal entries, a BE-matrix contains more information than a corresponding adjacency or connectivity matrix. Due to their algebraic properties [11,27] the BE-matrices are used in a great variety of computer programs for the deductive solution of chemical problems [26]. Thus, a BE-matrix-based documentation system can be directly interfaced with computer programs of the aforementioned type.

The chemical constitution of a molecule with n atoms can be represented by up to n! equivalent n x n BE-matrices, since the atoms can be indexed in up to n! distinct ways. For chemical documentation it is necessary to have a one-to-one correspondence of chemical compounds and their molecules, on the one hand, and the matrices which represent these molecules, on the other. To this end, for each compound one of the equivalent matrices is picked as the canonical representation of its molecules. This is achieved using algorithms by which the atoms are uniquely indexed, e. g. the Morgan algorithm [18], ASI [28], NOON [29], CANON [9,10]. Among these algorithms CANON is the only one which accounts for constitutional symmetries directly and in an adequate manner.

The CANONical indices of the atoms in a molecule indicate the sequential order of the of the atoms. The CANON procedure is illustrated by

3.1 Example: In a molecule, e. g. 1 , the atoms are first labelled (by letters here) in an arbitrary manner.

$$H^e - C^f - C^g - C^h - C^i$$

with substituents:
H^a, H^b, H^c, O^d (top)
H^j, Cl^k, Cl^l, $O^m - H^n$ (bottom)

1

These arbitrary labels (AL) serve to keep track of the identity of the individual atoms.

Next zeroth order atom indices (0.AI) are assigned to the atoms. The atoms with the highest atomic number are all given the index 1. Atoms with the next lowest atomic number receive the index = 2 etc., until the atoms with the lowest atomic number are reached. The nth descriptor (n.AD) of an atom is given by its n.AI, followed by the n.AI of its α-atoms (i. e. directly covalently connected neighboring atoms) in numerical order. Lexicographic ordering of the 0.AD gives the first order AI (1.AI). The 1.AI are used just as the 0.AI to find the first order AD (1.AD) and from them the 2.AI. This procedure continues until the number of different AI does not increase from iteration n to n + 1. If the molecule does not contain any rings, the n.AI are equal to the equivalence class indices (EI). Atoms with equal EI belong to the same equivalence class and are constitutionally equivalent. In the above example the process stops after two steps (see Table 8).

Table 8: The determination of the EI of 1 by the CANON algorithm

AL	O.AI	O.AD	1.AI	1.AD	2.AI	2.AD	EI
a	4	4:3	8	8:6	10	10:7	12:
b	4	4:3	8	8:4	9	9:5	11:
c	4	4:3	8	8:4	9	9:4	10:
d	2	2:3	2	2:5	2	2:6	3:
e	4	4:3	8	8:6	10	10:7	12:
f	3	3:3.4.4.4	6	6:4.8.8.8	7	7:5.10.10.10	8:
g	3	3:1.3.3.4	4	4:1.4.6.8	5	5:1.4.7.9	6:
h	3	3:1.3.3.4	4	4:1.4.5.8	4	4:1.5.6.9	5:
i	3	3:2.2.3	5	5:2.3.4	6	6:2.3.4	7:
j	4	4:3	8	8:6	10	10:7	12:
k	1	1:3	1	1:4	1	1:5	2:
l	1	1:3	1	1:4	1	1:4	1:
m	2	2:3.4	3	3:5.7	3	3:6.8	4:
n	4	4:2	7	7:3	8	8:3	9:

In order to generate the canonical indices (CI), one atom of the lowest ranking equivalence class which contains more than one atom is assigned an AI lower than all AI of the atoms in the same equivalence but otherwise not alterning the order of equivalence classes. Then new AD are constructed and the iterative procedure starts again. The whole procedure is executed until each equivalence class contains one atom. At this point the AI are termed the CI.

If a molecule contains rings, equal numbers of equivalence classes from iteration n to n + 1 is only necessary for having generated the EI. In these cases the ring descriptor (RD) is used in addition after the iteration stops to generate the EI. The length of the RD is given by the number of rings in the molecule, i. e., the maximal number of linearly independent cycles = 1 + (edges) - (vertices), according to the Euler - Poincaré theorem. The elements of the RD are given by the smallest rings the atoms considered belongs to.

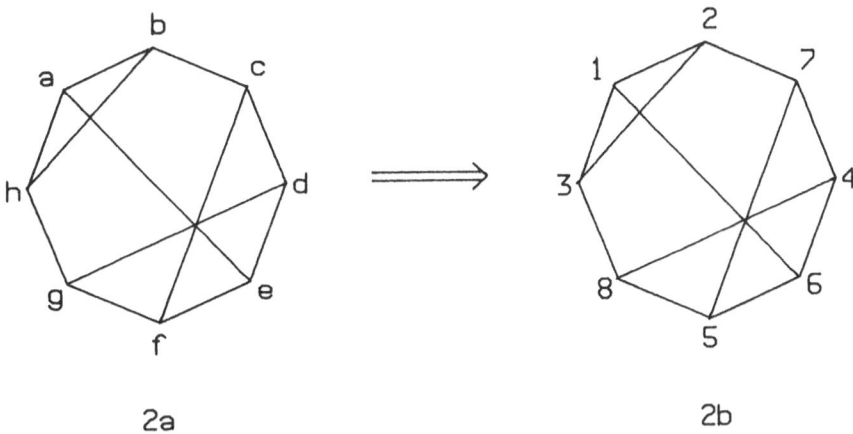

2a   2b

3.2 Example: All atoms in 2 are of the same kind, and for each of them the AD is 1.111. Since the hypothetical molecule contains rings, the RD must be constructed.

Table 9: The determination of the EI of 2

AL	RD	1.AI	1.AD	2.AI	2.AD	3.AI	EI
a	35555	1	1.112	1	1.112	1	1
b	35555	1	1.112	1	1.113	2	2
c	44555	2	2.122	3	3.144	4	4
d	44555	2	2.222	4	4.233	5	5
e	44555	2	2.112	2	2.144	3	3
f	44555	2	2.222	4	4.233	5	5
g	44555	2	2.122	3	3.144	4	4
h	35555	1	1.112	1	1.113	2	2

As an extra benefit available only with CANON the constitutional equivalencies of atoms can be detected while their indexing is being established. Indeed, any two atoms with the same EI are constitutionally equivalent.

The explicit detection of constitutionally equivalent atoms is not only essential for avoiding redundancies in documentation of the chemical constitution of molecules and for any efficient computer-assisted procedures for the solution of chemical problems [11], and is also absolutely indispensable for computer-assisted stereochemistry, including stereochemical documentation [26,30].

4. *Representation of the Stereochemical Features of Molecules*

Having considered the constitutional representation of molecules, we now consider the stereochemical features. To describe the stereochemical features, we shall rely on the notion of permutational isomerism. In combination with a representation of the chemical constitution of molecules by CANON and BE-matrices, the permutational description of the stereochemical features of molecules leads to a unified chemical documentation and nomenclature system. In this system the CANON indexing system provides the requisite interface between constitutional chemistry and stereochemistry.

4.1 *Molecular Skeleton and Set of Ligands*

The permutational treatment of stereochemistry begins with the conceptual dissection of a molecule into a set of ligands and a skeleton; the stereochemical features of the molecule are then described by the placement of the ligands on the skeletal sites.

It is particularly convenient that the skeleton should contain the central atoms of the asymmetric carbon atoms; those that are not constitutionally equivalent to other asymmetric carbon atoms [15,31] can then be regarded as stereogenic subunits. However, it is not always feasible to take these units as monocentric configurations, as, for example, in chiroinositol 3 [8] and its stereoisomers.

3

The possibility of such a dissection into monocentric configurations can easily be determined by using CANON. A carbon atom whose neighboring atoms are constitutionally nonequivalent is, for instance, detected as an ordinary asymmetric carbon atom (asymmetric carbon atoms with some stereoisomeric ligands are detected by other procedures). If all of the asymmetric carbon atoms are found to be independent (i. e. none of their central atoms are constitutionally equivalent according to their CANONical atomic EI), the configurational aspect of the given molecular system can be described in terms of asymmetric carbon atoms as local configurational subunits, and the conceptual dissection into skeletons and sets of ligands can be performed accordingly. If, however, constitutionally equivalent configurational subunits are detected, or if there are other reasons to represent the stereochemical features on the basis of polycentric molecular skeleton, another conceptual dissection into skeleton and ligands is required.

5. *Indexing Rules and Algorithms for Ligands*

There are essentially two procedures for establishing the indices of the ligands of permutation isomers. One is founded on the CIP rules, while the other relies on the CANON algorithm. Both correspond in essence to self-indexing systems, i. e. in both cases indexing is based on the chemical nature of the ligands.

5.1 *The CIP Rules*

Great progress in stereochemistry is due to two achievements first published in 1951: the determination of absolute configurations by Bijvoet, Peerdeman and Bommel [32] through a novel X-ray method, and the unambiguous description of configurations by the R,S-nomenclature, which is based on the CIP sequence rules for ligands [2-5]. In 1965 the unambiguous p,n-nomenclature of diastereoisomers [33] was derived from the R,S-nomenclature[*]. Probably the end of the uncertainty about whether or not all stereoformulas of chiral molecules must be replaced by their mirror images, stimulated the development of a configurational nomenclature that did not have the ambiguities of the classical D,L-nomenclature [6-8]. The D,L-nomenclature was based on a configurational correlation of the given molecule with D-glyceraldehyde as a reference molecule. But the exact configuration of D-glyceraldehyde was not known until 1951 [32], and also it was not possible in many cases to state clearly whether a given configuration resembles D-glyceraldehyde 7, or its enantiomer L-glyceraldehyde $\bar{7}$.

A further reason for the ambiguities of the D,L-nomenclature so well as the classical cis,trans-nomenclature [3] is that none of the classical stereochemical nomenclature systems contained clear-cut and universally applicable criteria by which a set of ligands could be ordered.

This was first recognized by Cahn and Ingold [2], who, together with Prelog [3-5], formulated the CIP rules for establishing sequential order in sets of ligands, in particular those of asymmetric carbon atoms. The CIP index of a ligand depends primarily on the atomic number of its α-atom, i. e. that atom of the ligand that is directly connected with the central atom by a covalent bond. The higher the atomic number of its α-atom, the

----

*) Recently Prelog and Seebach [23] have proposed replacement of the descriptors "p" and "n" by "lk" and "ul", because in 1966 the letter "P" has been reserved for "positive helicity" [4].

higher the sequential priority of the ligand, and the lower its sequential index. If the sequential priority of ligands cannot be decided by its α-atom, then its β-atoms are considered, then its γ-atoms etc. With the exeption of hydrogen, the coordination number of all atoms (e. g. atoms that participate in multiple bond systems) is brought to four by the addition of "duplicate atoms" according to the bond order of the multiple bonds and if necessary "saturated" with "phantom" atoms whose atomic number is zero [4]. There are further subrules concerning resonance structures, atomic mass numbers for isotopically different ligands and subrules for cis,trans-isomers and molecules with more than one asymmetric carbon atom. The CIP rules presently dominate stereochemical nomenclature, although their application is sometimes cumbersome, and they are by no means ideally suited to computer-assisted documentation.

The CIP rules have been critically examined by Hirschmann and Hanson [34] as well as by Schubert and Ugi [10]. The criticism voiced has concentrated on two points:

(1) Until the introduction of a recent revision [5], the CIP rules did not contain an explicit termination criterion, contrary to an explicit statement [5]. However, termination of the CIP procedure was implied in some parts of the 1966 paper on the CIP rules [4].

(2) The use of bond orders within the CIP rules is the source of cumbersome procedures in the case of multiple bond systems (requiring duplicate atoms, phantom atoms) and resonance systems (requiring finding the "standard" resonance structure).

All troubles that arise from π-electrons could be easily avoided by a simple revision. Moreover, the use of coordination numbers instead of bond orders within the CIP rule system, would yield a simple, easy-to-use system of rules for sequentially ordering the ligands of any configuration-rules which would be well-suited as a basis for computer-assisted chemical documentation and would be fully compatible with constitutional documentation [9,10]. In fact, the resulting indexing of ligands would coincide with that obtained by CANON (but without many of the builtin benefits of CANON, such as algorithmic simplicity, recognition of constitutional symmetry etc).

5.2 *The CANONical Ligand Indices*

An indexing of ligands which corresponds roughly to the CIP sequential order is obtained when the ligands are sequentially ordered according to the CANONICal equivalence class indices of their α-atoms, i. e. the atoms which are the immediate, covalently connected neighbors of the central atom of a monocentric configurational subunit of the skeleton. In the case of a polycentric skeleton, it seems advisable to treat the skeleton as a "superatom", whose coordination number is the number of ligands on the polycentric skeleton.

Example:

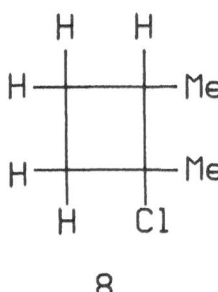

8

In 8 the α-atom of the methyl group next to the chlorine ligand would be assigned a lower EI than the α-atom of the other methyl group, although the methyl groups are indistinguishable as ligands.

Representation of configurations and other stereochemical features by the α-atoms of ligands was introduced in June 1973 at the NATO conference in Noordwijkerhout [19,20,28] on "computer representation and manipulation of chemical information". The use of the Morgan indices of the α-atoms was proposed by Wipke [19], and the use of the ASI indices of the α-atoms of ligands was proposed by Ugi et al. [28].

Indexing of the ligands of permutation isomers according to the sequence of the CANONical indices of the α-atoms of the ligands has the advantage that it connects the stereochemical aspect with the constitutional aspect. It takes constitutional symmetries into account, and the ligand indices correspond in most cases to the presently used version of the CIP rules [4,5]. The indexing of ligands by CANON and by the CIP rules will become identical, whenever the CIP rules undergo their remaining desirable major revisions, namely the replacement of the bond order based subrules by a rule relying on coordination numbers. An extra bonus of

CANON-based ligand indexing concerns those cases requiring otherwise primed indices [35]. In such cases CANON can be used in a straightforward manner to yield a satisfactory ligand indexing.

6. *The Reference Isomer*

With n chemically distinguishable ligands and a rigid molecular skeleton a family of n! distinct molecular models can be generated. One of these models is taken to be the reference model E of this family. Some members of the family are interconvertible with E by rotations of the whole molecule. These are chemically identical with E and belong to the reference isomer X. The ligand permutations whose actions interconvert the distinct molecular models of the reference model E form its chemical identity group $S_E = S_X$.

Any ligand permutation not belonging to the chemical identity group S_X of the reference isomer X converts its models into models for one of its permutation isomers.

6.1 *Skeletal Classes and their Reference Isomer*

For many applications of the present theory it is necessary to choose a reference model E or a reference isomer X within its family of models or permutation isomers.

In principle any member of a family may serve as a reference, but it is advantageous to define a reference system for each family which is con-

184

sidered repeatedly. For nomenclature and documentation purposes fixed reference systems are a prerequisite.

Let us assume a reference isomer with n different ligands [12]. In this reference isomer, the ligands are represented by numbers. With a given set of ligands these numbers are assigned to the individual ligands. As long as the chemical nature of the ligands is not specified, such a reference molecule may be used as the reference isomer of any family of permutation isomers which has the given skeleton and n distinguishable ligands. Thus, one such reference molecule may be used for a whole skeletal class, i. e. molecules that have in common a specified skeleton. The particular family of permutation isomers is determined by the chemical nature of the ligands.

6.2 *Choice of Reference Isomers*

In the first paper on permutation isomerism and permutational nomen-clature [12] the reference isomer of the most widely encountered skeletal classes were defined in a more or less arbitrary manner to be e. g. 9 - 14 (see also I,1).

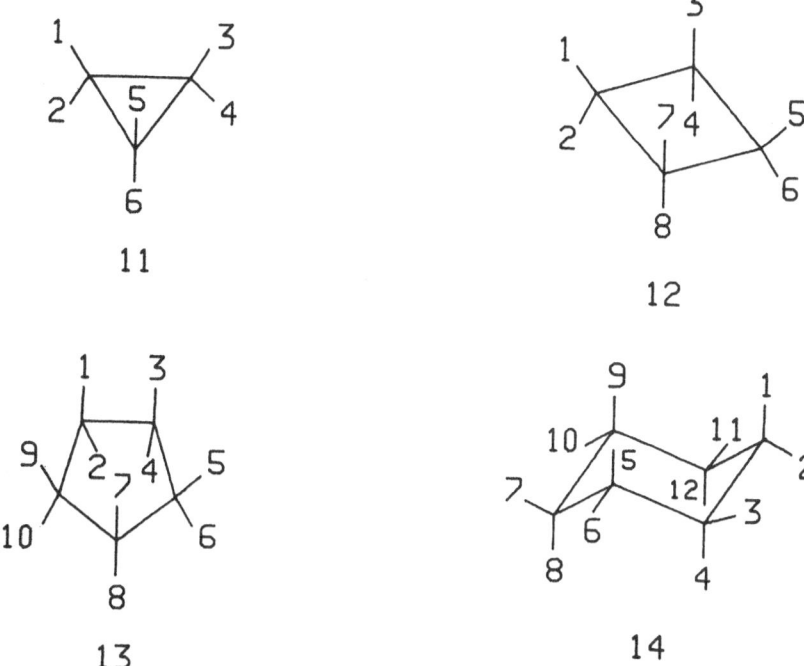

One could give some sets of rules for choosing the reference isomer.   However,   these would probably be very cumbersome, if they are to cover all of
stereochemistry. A systematic approach to the selection of a reference isomer from a pair of enantiomers is given in 6.4.

6.3   Ordering and Selection Rules for Sets of Permutations

Since with a given reference isomer each of its permutation isomers is
equivalently represented by elements of a coset   (see 7.1)   or double coset
(see 7.2),   rules are needed for selecting one permutation from   a   set   of
permutations, as the permutational descriptor of a molecule.   We have found
that the following rules serve the purpose particularly well:

Let A be a linearly ordered set of n objects, e. g. the integers 1,2,...,n written in their natural order. By a permutation of A is meant simply a linear rearrangement of the given objects, i. e. placement of the integers 1,...,n in some other order.

The customary cyclic notation for permutations indicates only which objects in the linear order 1,2,...,n are to be interchanged. For our purposes we work directly with the resulting rearrangement that the given permutation specifies.

Thus, to each permutation in the customary cyclic notation is associated a definite rearrangement of the objects in A, which we call the string form of this permutation. We order any set of permutations of A by the lexicographic ordering of their string forms, i. e. the rearrangements that they represent [36].

Example: In S_4, consider the permutations (14),(23) and (13)(24); they represent the rearrangements {4231}, {1324} and {3412} respectively, therefore (23)<(13)(24)<(14).

6.4 Chiral Reference System

The following system of non-intersecting skew lines may serve as a universal reference system for chiral and helical objects [4,5,38].

To fix the notation, we assume given two oriented lines, g_1 and g_2, in 3-space which do not intersect and are not parallel. The shortest distance between g_1 and g_2 is unique and is given by a line d which is perpendicular to both g_1 and g_2. We now choose a point P on d not on the segment joining g_1 and g_2. When viewed from P, one of the lines

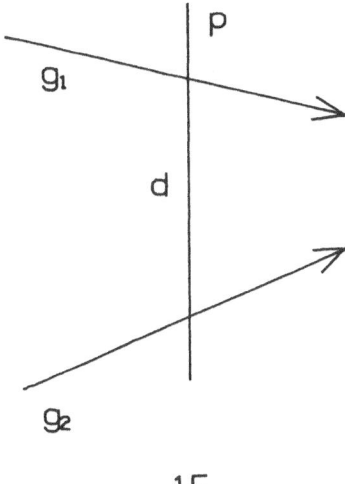

15

g_1, g_2 is seen to be in 'front' and the other 'behind'. The lines are provided with arrows to indicate their direction. Let α be the smallest angle by which the 'top' line must be rotated to bring its arrows into coincidence with that of the 'back' line (this angle is always $<180°$). The system is called right-handed, or an R-system if α is described by a clockwise rotation of the 'top' line.

Remark: There is no difference from which side along d we view these lines: in both cases, the angle and sense of rotation of the lines we see on 'top' are the same.

For a given chiral molecule, the chiral arrangement of skew oriented lines is determined as follows: the first line goes from ligand 1 to ligand 3, and the second line goes from ligand 2 to the smallest ligand k for which the line 1 → 3 is skew to 2 → k (such a line must exist, else the molecule is flat and therefore not chiral).

From a pair of enantiomers, that one is selected for which the skew oriented lines form an R-system. This selection of the reference isomer from an enantiomer pair has been chosen to be in accordance with the traditional R,S-nomenclature by associating an asymmetric carbon atom having R-configuration with an R-system of oriented skew lines. Note that a chiral system of skew unoriented lines is used as a chiral reference for the Λ,Δ-nomenclature [38] of complexes with a hexacoordinate skeleton which has O_h-symmetry.

7. *Permutational Descriptors*

7.1 *Permutation Isomers with Chemically Distinguishable Ligands*

With a given reference isomer (see II,1), its permutation isomers are represented by the left cosets of the chemical identity group of the reference isomer. One could use these cosets as the descriptors of the individual permutation isomers. However, any member of a coset may serve equally well as the descriptor of the permutation isomer [12]. To have uniqueness of the permutational descriptors, the first member of each coset is selected as its representative according to the rules given in VIII,6.3.

This procedure has a substantial disadvantage, namely that enantiomers are not recognized as such by their descriptors. Therefore, we propose a somewhat more practical but also more complicated procedure. First, the

189

given permutation isomer is described up to enantiomers by the respective
left coset of the racemate group of the reference isomer and its enantiomer
(see IV,5). Then a permutation is chosen from such a coset as above, and
it is determined whether the action of this ligand permutation leads to the
given isomer from the reference isomer or from its enantiomer. If this permutation leads from the reference isomer to the given isomer, then it is
underlined and directly used as the descriptor of the given molecule. If,
however, the considered isomer is obtained from the enantiomer of the
reference isomer by the action of this permutation, it is labelled with a
bar, and is then used as the permutational descriptor. Descriptors without
bars are used for achiral molecules.

Example: With 16 as the reference isomer X, the permutation (15) belongs to
17 and indicates that an exchange of ligands 1 and 5 leads from 16 to 17.
The permutation (152) belongs to $\overline{17}$, the enantiomer of 17, and indicates
that $\overline{17}$ is obtained from 16 through the action of (152), or by permuting
the ligands 1 and 5 and subsequent conversion into the enantiomer by
exchanging 2 and 5, represented by (25)(15).

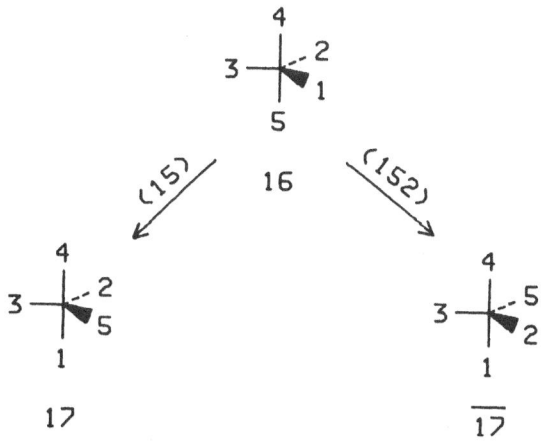

The permutational descriptors of 17 and $\overline{17}$ are $(\overline{14532})$ and $(\underline{14532})$. These

descriptors are found as follows:

Starting from

$$R_{16} = S_{16} \cup \overline{S}_{16} = \{e,(123),(132),(12)(45),(13)(45),(23)(45)\}$$
$$\cup \{(12),(13),(23),(45)(123)(45),(132)(45)\};$$

or converted to string notation:

$$R_{16} = \{12345, \ 23145, \ 31245, \ 21354, \ 32154, \ 13254\}$$
$$\cup \{21345, \ 32145, \ 13245, \ 12354, \ 31254, \ 23154\}$$

we obtain the coset

$$(15)R_{16} = (15)S_{16} \cup (15)\overline{S}_{16}$$
$$= \{(15),(1523),(1532),(1452),(1453),(145)(23)\}$$
$$\cup \{(152),(153),(15)(23),(145),(14523),(14532)\}$$

and convert that to string notation:

$$(15)R_{16} = \{52341, \ 53142, \ 51243, \ 41352, \ 42153, \ 43251\}$$
$$\cup \{51342, \ 52143, \ 53241, \ 42351, \ 43152, \ 41253\}$$

We now choose the lexicographically lowest permutation in $(15)R_{16}$,

namely $41253 \approx (14532)$, as the descriptor of 17. Since $(14532) \in (15)\overline{S}_{16}$,

we use $(\overline{14532})$ as the descriptor of 17 and $(\underline{14532})$ as the descriptor of the

enantiomer $\overline{17}$.

A modification of this procedure simplifies the search somewhat. One

selects a permutation which converts the reference isomer 16 into 17, say

$\lambda = (15) \approx 52341$. The lowest permutation in $(15)R_{16}$ would result from

members of R_{16} which remove the "5" from the beginning of 52341, i. e. the

permutations acting on 5. In R_{16} these are the permutations containing the

cycle (45), whose action on 52341 results in strings 4..5.; since

((123)(45) leads to 41253, this permutations in cycle notation

(15)(123)(45) = (14523) is selected to represent 17.

The latter procedure has not only particular advantages in computer assisted descriptor assignment, but also for non-automated descriptor search. With some experience, one glance at R_X in string notation suffices to find within R_X the permutation which gives the lexicografically lowest member of $^\lambda R_X$.

7.2 Molecules with some indistinguishable Ligands

If the ligands are not all distinguishable (see VI,2), the equivalency of ligands is represented by a stabilizer group Σ and the permutation isomers correspond to the double cosets $\Sigma\lambda S_X$. The descriptors are picked as in VIII,7.1; the double cosets instead of the cosets λS_X are used without a label if achiral and with a configuration-dependent label if chiral. Molecules with subsets of indistinguishable ligands are formally obtained from molecules with all ligands distinguishable by ligand substitutions.

Example: The molecules 17, 18, 19 and 20 are obtained from their reference isomer 16 by the action of the ligand permutations (15),(125),(15)(34) and (125)(34). The aforementioned permutation isomers of 16 are all converted into 21 when ligand 2 is replaced by 1, and 4 is replaced by 3.

Note that the stabilizer group of 21

$$\Sigma = \{e,(12),(34),(12)(34)\}$$

intersects with the cosets $(15)S_{16}$, $(125)S_{16}$, $(15)(34)S_{16}$ and $(125)(34)S_{16}$

(see VII, Table) which represent 17 - 20. The union of these cosets is the double coset $\Sigma(15)S_{16}$ which represents 21 (see V,2).

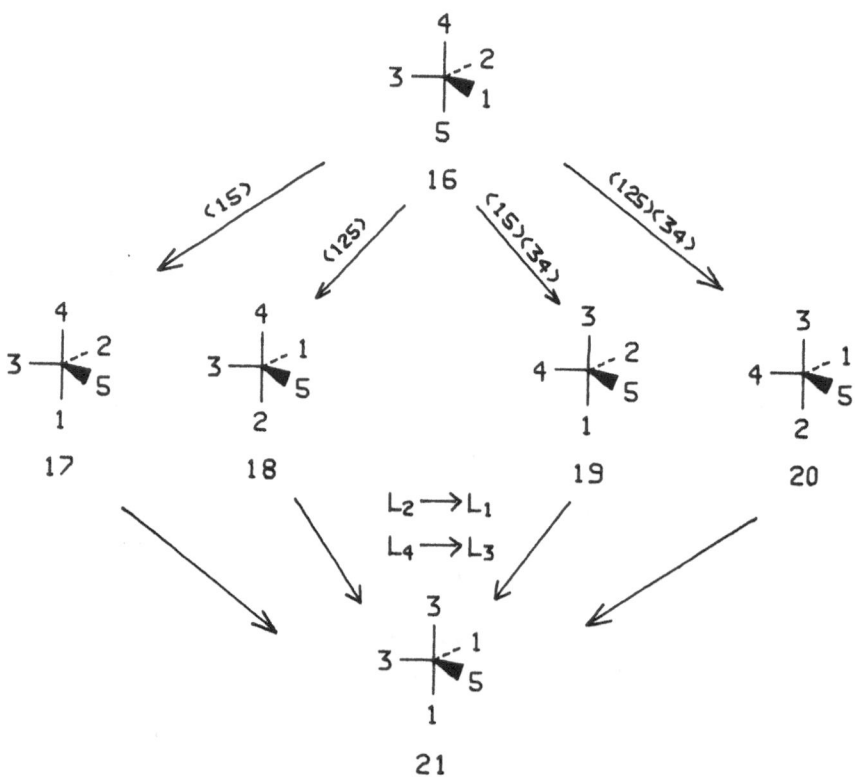

The lexicographically lowest among the members of the double coset $\Sigma(15)R_{16}$ is part of the descriptor of 21, namely $(2453) \approx 14253$. Since 21 is chiral, we must determine whether $(\underline{2453})$ or $(\overline{2453})$ is the descriptor of 21. Since (125) and (2453) do not belong to the same coset of S_{16}, but to "enantiomeric cosets", the descriptor of 21 is $(\overline{2453})$. The descriptor of 21 can also be found by determination of the lexicographically lowest permutation in each of the $\lambda R_{16} \subset \Sigma(15)R_{16}$ by the modified procedure used in the preceding example. The representation of these cosets are

$41253 \approx (14532) \in (15)R_{16}$, $14253 \approx (2453) \in (125)R_{16}$, $41523 \approx (142)(35)$

ϵ (15)(34)R_{16} and 14523 $\approx$ (24)(35) ϵ (125)(34)R_{16}, (2453) being the lowest of these.

The reconstruction of 21 from its descriptor is straightforward. The permutation (2453) converts 16, the reference isomer, into $\overline{18}$ which is converted into $\overline{21}$ by the substitutions $L_2 \to L_1$ and $L_4 \to L_3$. Thus the molecule belonging to (2453) is $\overline{21}$, and its enantiomer 21 is described by $(\overline{2453})$.

This seemingly cumbersome two-step procedure is necessary, because ligand permutations on molecules with some indistinguishable ligands can lead to ambiguities, as is seen in the case

194

References

[1] J. Brandt, J. Bauer, R. M. Frank and A. von Scholley, Chemica Scripta 18, 53 (1981).

[2] R.S. Cahn and C. K. Ingold, J. Chem. Soc. 1951, 612.

[3] R. S. Cahn, C. K. Ingold and V. Prelog, Experientia 12, 81 (1956).

[4] R. S. Cahn, C. K. Ingold and V. Prelog, Angew. Chem. 78, 413 (1966).

[5] V. Prelog and G. Helmchen, Angew. Chem. 94, 614 (1982); Angew. Chem. Int. Ed. 21, 567 (1982); see also: E. F. Meyer, J. Comput. Chem. 1, 229 (1980).

[6] E. L. Eliel, "Stereochemistry of Carbon Compounds", Mc Graw-Hill, New York 1962; K. Mislow, "Introduction to Stereochemistry", Benjamin, New York 1965; H. B. Kagan, "Organische Stereochemie", G. Thieme Verlag, Stuttgart 1977; M. Nógradi, "Stereochemistry", Pergamon Press, Oxford 1981.

[7] W. Bähr and H. Theobald, "Organische Stereochemie", Springer Verlag, Heidelberg 1973.

[8] J. Dale, "Stereochemie und Konformationsanalyse", Translation: A. Senning, Verlag Chemie, Weinheim 1978.

[9] W. Schubert and I. Ugi, J. Amer. Chem. Soc. 100, 37 (1978)

[10] W. Schubert and I. Ugi, Chimia 33, 183 (1979).

[11] J. Dugundji and I. Ugi, Top. Curr. Chem. 39, 19 (1973).

[12] I. Ugi, D. Marquarding, H. Klusacek, G. Gokel and P. Gillespie, Angew. Chem. 82, 741 (1970); Angew. Chem. Int. Ed. 9, 703 (1970).

[13] P. Gillespie, P. Hoffmann, H. Klusacek, D. Marquarding, S. Pfohl, F. Ramirez, E. A. Tsolis and I. Ugi, Angew. Chem. 83, 691 (1971); Angew. Chem. Int. Ed. 10, 687 (1971).

[14] J. Dugundji, P. Gillespie, D. Marquarding, I. Ugi and F. Ramirez, in "Chemical Applications of Graph Theory", ed.: A. T. Balaban, Academic Press, London 1976, p. 107.

[15] J. Gasteiger, P. Gillespie, D. Marquarding and I. Ugi, Top. Curr. Chem. 48, 1 (1974).

[16] R. C. Weast and M. J. Astle, ed.: "CRC Handbook of Chemistry and Physics", CRC Press, Boca Raton, Fla. 1979, p. C-1; D. Hellwinkel, "Die systematische Nomenklatur der organischen Chemie", Springer, Heidelberg 1982.

[17] a) W. J. Wiswesser, Chem. Eng. News 30, 3523 (1952); " Lineformula Chemical Notation", Crowell Co., New York 1954; E. G. Smith, "The

Wiswesser Line-formula Chemical Notation", Mc Graw-Hill, New York 1968; R. Fugmann, H. Nickelsen, I. Nickelsen and J. H. Winter, Angew. Chem. 82, 611 (1970); Angew. Chem. Int. Ed. 9, 589 (1970); b) J. E. Ash and E. Hyde, "Chemical Information Systems", Wiley & Sons, New York 1975.

[18] H. L. Morgan, J. Chem. Doc. 5, 107 (1965); see also: W. C. Herndon and J. E. Leonard, Inorganic Chemistry 22, 554 (1983).

[19] W. T. Wipke and T. M. Dyott, J. Amer. Chem. Soc. 96, a) 4825, b) 4834 (1974).

[20] W. T. Wipke in "Computer Representation and Manipulation of Chemical Information", ed.: W. T. Wipke, S. R. Heller, R. J. Feldman and E. Hyde, Wiley & Sons, New York 1974, p. 147.

[21] J. Weyer, Angew. Chem. 86, 604 (1974); Angew. Chem. Int. Ed. 13, 591 (1974).

[22] Lord Kelvin, "Baltimore Lectures", Clay &Sons, London 1904.

[23] D. Seebach and V. Prelog, Angew. Chem. 94, 696 (1982), Angew. Chem. Int. Ed. 21, 654 (1982); see also: D. Hoppe, Nachr. Chem. Tech. Lab. 30, 1030 (1982); K. Mislow and J. Siegel, J. Amer. Chem. Soc. (in press).

[24] O. Aschan, Ber. dtsch. chem. Ges. 35, 3389 (1902).

[25] L. Spialter, J. Chem. Doc. 4, 261 (1964).

[26] I. Ugi, J. Bauer, J. Brandt, J. Friedrich, J. Gasteiger, C. Jochum and W. Schubert, Angew. Chem. 91, 99 (1979), Angew. Chem. Int. Ed. 18, 111 (1979).

[27] C. Jochum, J. Gasteiger, I. Ugi and J. Dugundji, Z. Naturforsch. 37b, 1205 (1982).

[28] J. Blair, J. Gasteiger, C. Gillespie, P. D. Gillespie and I. Ugi, in "Computer Representation and Manipulation of Chemical Information", ed.: W. T. Wipke, S. R. Heller, R. J. Feldman and E. Hyde, J. Wiley & Sons, New York 1974, p. 129; Tetrahedron 30, 1845 (1974).

[29] C. Jochum and J. Gasteiger, J. Chem. Inf. Comput. Sci. 17, 113 (1977).

[30] H. W. Davis, "Computer Representation of the Stereochemistry of Organic Molecules", Birkhäuser ISR23, Basel 1976.

[31] J. Dugundji, R. Kopp, D. Marquarding and I. Ugi, Top. Curr. Chem., 75, 165 (1978).

[32] J. M. Bijvoet, A. F. Peerdeman and A. J. van Bommel, Nature (London) 168, 271 (1951).

[33] I. Ugi, Z. Naturforsch. 20B, 405 (1965); Nachr. Chem. Tech. Lab. 31, 276 (1983).

[34] H. Hirschmann and K. R. Hanson, Tetrahedron 30, 3649 (1974).

[35] M. F. Brown, B. R. Cook and T. E. Sloan, Inorg. Chem. 14, 1273 (1975).

[36] This approach was brought to our attention by J. Bauer who used the string form of permutations in IGOR [37].

[37] J. Bauer, Dissertation, Technische Universität München 1982; J. Bauer and I. Ugi, J. Chem. Res. 1982, (S) 298, M 3101, 3201.

[38] A pair of skew, non-intersecting, non-oriented lines has been introduced as a chiral reference system for the ΔΛ-nomenclature [39].

[39] Definition by the IUPAC-CNIC: Inorg. Chem. 9, 1 (1970); see also: M. Brorson, T. Damhus and C. Schäffer, Inorg. Chem. 22, 1569 (1983).

It is assumed that the reader has some knowledge of basic group theory. The purpose of this appendix is to compile the notation and terminology used in the text, and to recall the group-theoretic results that are needed; proofs will be given only for those results that are not immediately available in standard texts [A, B, C, D].

We use the customary symbols for set-operations; in particular, $\{x|P\}$ will denote the set of all objects x having the property P, and O will denote the empty set. If A, B are two sets, a function $f:A \to B$ is a rule that assigns to each element $a \in A$ an element $f(a) \in B$; the function f is also denoted by $a \mapsto f(a)$ whenever we want to display the rule by which the function is defined. A function $f:A \to B$ is also called a map of A into B, and $f(a) \in B$ is called the image of $a \in A$; the image of a subset $S \subset A$ is the set $f(S) = \{f(s)|s \in S\} \subset B$. The map $f:A \to B$ is injective if $f(a) = f(c)$ always implies $a = c$; it is surjective if $f(A) = B$, and bijective (or one-to-one) if it is both injective and surjective. If $f:A \to B$ and $g:B \to C$ are two maps, their composition $g \circ f:A \to C$ is the function defined by the rule $a \mapsto g[f(a)]$; note explicitly that in the symbol $g \circ f$, the order in which the mappings are performed is read from right to left.

Recall that a group G is a set together with a binary operation (called multiplication) that assigns to each ordered pair (a,b) of elements of G an element $a \cdot b \in G$ and satisfies the three conditions.

1. (Associativity) $(a \cdot b) \cdot c = a \cdot (b \cdot c)$ for all $a,b,c \in G$

2. (Existence of a unit)   There is some element e ∈ G   (called the unit)

    such that   e·a = a·e = a   for each a ∈ G.

3. (Existence of inverses)   For each a ∈ G   there is an element (written

    a^{-1} and called the inverse of a)   such that $a·a^{-1} = a^{-1}·a = e$.

The unit e of a group is unique, as is the inverse a^{-1} of each a ∈ G;   the

inverse of the product a·b   is $b^{-1}·a^{-1}$.

    A group is denoted by (G,·)   whenever we wish to display the group

operation being used;   a group (G,·) is called abelian if a·b = b·a for all

a,b ∈ G.

    Whenever the group operation being used is clear, we denote the group

(G,·)   simply by G, and the product a·b by ab. The set of elements in a

group G may be finite or infinite. If the set is finite, G is called a

finite group, and the number |G| of elements is called the order of the

group. If the set is infinite, we say that the order of G is infinite and

write |G| = ∞.

1. Subgroups   Let (G,·) be a group. A subset H ⊂ G such that (H,·) is

itself a group is called a subgroup of (G,·).

1.1   H ⊂ G is a subgroup if and only if both   (i) Whenever a,b ∈ H   then

also a·b ∈ H and (ii) Whenever a ∈ H, then also a^{-1} ∈ H.

Clearly, {e} and G itself are subgroups of G, and every subgroup of G

contains e. It is very useful to observe that, if G is a finite group then

the condition (i) of 1.1 alone characterizes its subgroups:

1.2   Let G be a finite group. A subset $H \subset G$ is a subgroup if and only if whenever $a, b \in H$ then also $a \cdot b \in H$.

Proof   "If":   According to 1.1, we need only show that if $a \in H$, then $a^{-1} \in H$. Proceeding inductively, we find from $a \in H$ that $a^2 = a \cdot a \in H$, $a^3 = a^2 \cdot a \in H$, ... and therefore that all powers a^n belong to H. Since $|H|$ is finite, the a^n can not be all different, so $a^n = a^m$ for some $n \neq m$. Assuming $n > m$, we therefore have $e = a^{n-m}$, where $k = n - m$ is a positive integer. Since $e = a^k = a \cdot a^{k-1} = a^{k-1} \cdot a$, this shows $a^{k-1} = a^{-1}$ and there-fore $a^{-1} \in H$. The "only if" is trivial.

An important property of subgroups is

1.3   The intersection of any family of subgroups of G is also a subgroup of G.

If M is a non-empty subset of the group G, the intersection of all subgroups containing M is denoted by $\langle M \rangle$ and called the subgroup generated by M. In particular, $\langle M \rangle$ is unique and is the "smallest subgroup of G containing M", in the sense that $\langle M \rangle$ is contained in every subgroup that contains the set M. The elements of $\langle M \rangle$ can be described directly in terms of M; for finite groups (the case of main interest in the text) this is given in

1.4   Let G be a finite group, and M a subset of G. Then $\langle M \rangle$ is the set of all distinct finite products $a_1 \cdot a_2 \cdots \cdot a_s$ where each $a_i \in M$.

As an immediate consequence

1.5 If each member of M commutes with each member of a subset $L \subset G$, then so also will all members of $\langle M \rangle$.

A subgroup $H \subset G$ generated by a single element $g \in G$ (or, more generally, any group generated by a single element g) is called a cyclic group, and g the generator of that group. The order of $\langle g \rangle$ is called the order of the element g. Every element of a group G has an order ≥ 1, and e is the only element of order 1. If $|\langle g \rangle|$ is finite, its order is the smallest integer $n \geq 1$ such that $g^n = e$, and $\langle g \rangle$ consists of the distinct elements $\{e, g, g^2, \ldots, g^{n-1}\}$. If the order of g is n, then for any positive integer k, the order of g^k is $n/(n,k)$, where (n,k) is the greatest common divisor of n and k. Any cyclic group is necessarily abelian; and the only abelian groups G that have no subgroups other than $\{e\}$ and G are the cyclic groups of prime order.

In contrast to 1.3, the union of subgroups of G may not be a subgroup of G; the subgroup generated by the union of the subgroups $H_1, \ldots, H_n$ of G is denoted by $\langle H_1, \ldots, H_n \rangle$; it is the smallest subgroup of G that contains $\bigcup_1^n H_i$.

2. Cosets Given any two subsets M,T of a group $(G, \cdot)$ the set $\{m \cdot t \mid m \in M, t \in T\}$ of all products, where the first term belongs to M and the second belongs to T is denoted by $M \cdot T$. If G is not abelian, the set $M \cdot T$ may differ from the set $T \cdot M$, and even when M,T are both subgroups the set $M \cdot T$ may not be a subgroup.

If H is a subgroup of G and a is any element of G, the set a·H is called a left coset of H in G; clearly, e·H = H so that H itself is always a left coset of H in G. The elements of G belonging to any given left coset aH can be described more conveniently by

2.1   The left coset $aH = \{g \in G \mid a^{-1}g \in H\}$. Moreover, bH = aH for each b ∈ aH.

The basic property of the family $\{aH \mid a \in G\}$ of all the left cosets of H is

2.2   Each a ∈ G belongs to at least one left coset (for example, aH), and any two left cosets of H in G are either identical or disjoint. Therefore G is the pairwise disjoint union of all the distinct cosets of H in G. Moreover, if |H| < ∞, then each left coset contains precisely |H| distinct elements.

Any element of a coset aH is called a representative of aH; a system of representatives, one from each of the distinct cosets, is called a left transversal of H in G. Thus, for each left transversal $T = \{t_1, \ldots, t_n\}$ of H in G, we have that G = T·H and that G is the pairwise disjoint union

$$G = \bigcup_1^n t_i H.$$

The number of distinct left cosets of H in G is called the index of H in G, and is denoted by [G:H]. From 2.2 we find

2.3   If G is a finite group and H a subgroup, then [G:H] = |G|/|H|.

In particular, the order of any subgroup (and therefore the order of any element of G) is always a divisor of $|G|$.

As an extension of 2.3,

2.4 If $K \subset H \subset G$ we have two subgroups of a finite group G, then $[G:K] = [G:H] \cdot [H:K]$. Indeed, if $\{a_i \mid 1 \le i \le r\}$ is a left transversal of K in H, and $\{b_j \mid 1 \le j \le s\}$ is a left transversal of H in G, then $\{b_j \cdot a_i \mid 1 \le j \le s, 1 \le i \le r\}$ is a left transversal of K in G.

We now consider products $H \cdot K$ where H and K are subgroups of G. The main result is

2.5 Let K and H be any two subgroups of a finite group G, and let $\{h_1, \ldots, h_s\}$ be a left transversal of $H \cap K$ in H. Then the set HK is the pairwise disjoint union $HK = \overset{s}{\underset{1}{\cup}} h_i K$ and, in particular $|HK| = |H| \cdot |K| / |H \cap K|$.

We have remarked that HK may not be a subgroup of G, in fact

2.6 Let K, H be any two subgroups of a group G. A necessary and sufficient condition that HK be a subgroup is that HK = KH. Whenever HK is a subgroup, it is the subgroup $\langle H, K \rangle$ generated by H and K.

It follows from this that if G is abelian, then the product HK of any two subgroups will be a subgroup of G. Another very useful result involving products of subgroups is

2.7   (Dedekind's rule)   Let A ⊂ B be two subgroups of G, and let Q be any subgroup. If A ⊂ B ⊂ A·Q, then B = A·(B ∩ Q).

Because this is used frequently, we supply the simple proof: First note that since A ⊂ B and B ∩ Q ⊂ B, we have A·(B ∩ Q) ⊂ B. We now show B ⊂ A·(B ∩ Q). Let b ∈ B; since b ∈ A·Q we have b = a·g for some a ∈ A and g ∈ Q; now g = a⁻¹b and because A ⊂ B this shows g ∈ B, so g ∈ B ∩ Q and b ∈ A·(B ∩ Q). This completes the proof. Note that, if G is a finite group, it follows from 2.5 that

$$|B| = |A| \cdot \frac{|B \cap Q|}{|A \cap Q|} .$$

Let H be a subgroup of (G,·); by a right coset of H in G is meant a set Ha for some a ∈ G. A right coset of H in G is in general different from any left coset of H in G; however, there is a simple relation between them:

2.8   The inverses of the elements in any right coset Ha   form exactly the left coset a⁻¹H.

2.9   If a right coset Ha happens to be also a left coset of H, then it is necessarily the left coset aH.

Arguments using left cosets are formally applicable using right cosets; in particular, Ha = {g ∈ G | ga⁻¹ ∈ H}. The result 2.2 also holds, yielding a pairwise disjoint decomposition of G by the right cosets of H and it follows from this analog of 2.2 that there are as many distinct right cosets of H as there are distinct left cosets of H; the index of H in G computed by using right cosets is therefore the same as that computed using left cosets.

The coset concept can be generalized. Let H and K be any two subgroups of a group G; for each a ∈ G, the set HaK is called an (H,K)-double coset. It is easy to see that if b ∈ HaK, then HbK = HaK.

2.10 Each a ∈ G belongs to at least one (H,K)-double coset (for example, HaK) and any two (H,K)-double cosets are either identical or disjoint. Therefore G is the pairwise disjoint union of all the distinct (H,K)-double cosets.

Distinct (H,K)-double cosets may contain different numbers of elements; in fact

2.11 Let H,K be two subgroups of a finite group G, and HaK any double coset. Then

(a) $|HaK| = |H| \cdot |K| \, / \, |Ha \cap aK|$

(b) HaK is expressible as the (pairwise disjoint) union of $|K| \, / \, |Ha \cap aK|$ right cosets of H, and also as the union $|H| \, / \, |Ha \cap aK|$ left cosets of K.

3. **Normal subgroups** Let (G,·) be a group. A subgroup N ⊂ G is called normal in G (written N ◁ G) if each left coset of N is also a right coset of N. According to 2.9, each coset aN will then coincide with the coset Na, giving the convenient

3.1   The subgroup $N \subset G$ is normal in G if and only if $a^{-1}Na = N$ for each $a \in G$.

Clearly, $\{e\}$ and G are always normal subgroups of G; a group G that contains no normal subgroup other than $\{e\}$ and G is called a simple group.

As an example of a normal subgroup

3.2   If a subgroup $H \subset G$ has index 2 in G, then $H \triangleleft G$.

This is used so frequently that we include a proof: Chosing an a not in H, we have $G = H \cup aH = H \cup Ha$, therefore $aH = G-H = Ha$, so H is normal in G.

It is important to note that if $K \subset H \subset G$ are subgroups with $K \triangleleft H$ and $H \triangleleft G$, then it is not necessarily true that $K \triangleleft G$. However, if $H \triangleleft G$ and if K is any subgroup, then $H \cap K$ is normal in K; moreover,

3.3   If $N_1, \ldots N_s$ are normal subgroups of G, then $\overset{s}{\underset{1}{\cap}} N_i \triangleleft G$ and also $\langle N_1, \ldots, N_s \rangle \triangleleft G$.

It is clear that every subgroup of an abelian group G is normal in G. One useful property of normal subgroups is

3.4   Let $N \triangleleft G$ and $S \subset G$ be any subset. Then $N \cdot S = S \cdot N$.

From this and 2.6 follows

3.5   Let $N \triangleleft G$ and let H be any subgroup of G. Then $N \cdot H$ is a subgroup of G; in fact, $N \cdot H$ is the subgroup $\langle N, H \rangle$ generated by N and H; moreover, $N \cdot H = H \cdot N$.

The most important property of a normal subgroup $N \triangleleft G$ is that the product $(Na) \cdot (Nb)$ of any two cosets is precisely the coset Nab; with the binary operation $(Na) \cdot (Nb) = (Nab)$, the set of all distinct cosets of N can be shown to form a group, called the factor group of G by N and denoted G/N. The identity of the group G/N is the coset (N) and the inverse of $(Na) \in G/N$ is the coset (Na^{-1}).

Let M be any subset of a group G. Then $N_G(M) = \{a \in G \mid a^{-1}Ma = M\}$ is easily seen to be a subgroup of G; it is called the normalizer of M in G. Clearly, the group $N_G(M)$ may not contain the entire set M; but

3.6   If H is any subgroup of G, then $H \subset N_G(H)$ and, in fact, $N_G(H)$ can be characterized as the largest subgroup of G that contains H as a normal subgroup.

It is, of course, quite possible that $N_G(H) = H$; and H is normal in G if and only if $N_G(H) = G$. We remark that if K, H are two subgroups and $K \subset H$, it is not necessarily true that any one of $N_G(H)$, $N_G(K)$ is a subset of the other.

As a more restrictive property on a given subset $M \subset G$, we have that $C_G(M) = \{a \in G \mid a^{-1}ma = m$ for each $m \in M\}$ is also a subgroup of G, called the centralizer of M in G. In fact, $C_G(M)$ is the largest subgroup of G with the property that $am = ma$ for each $m \in M$. It is easy to see that $C_G(M) < N_G(M)$ and that if M consists of a single element m, then $C_G(m) = N_G(m)$.

Let H be a subgroup of G. The core of H in G is the largest subgroup of H that is normal in G; it is denoted by $core_G(H)$.

3.7 If H is any subgroup of G, then $core_G(H) = \cap \{gHg^{-1} \mid g \in G\}$.

   The number of terms in the intersection can be reduced by choosing a left transversal T of H in G; then $core_G(H) = \cap \{tHt^{-1} \mid t \in T\}$.

4. Conjugates   Let M be any subset of a group G. For each $a \in G$, the set $a^{-1}Ma$ is called a conjugate of M. Any two conjugates of M contain the same number of elements.

   If $H \subset G$ is any subgroup of G, then each conjugate $a^{-1}Ha$ is also a subgroup of G, and the set $\{a^{-1}Ha \mid a \in G\}$ of subgroups is called the conjugacy class of the subgroup H. It is of importance to know how many distinct groups a conjugacy class contains:

4.1   Let G be a finite group, and $H \subset G$ a subgroup. The number of dis-tinct groups in the conjugacy class of H is $[G:N_G(H)]$.

   For each fixed $m \in G$, the set $\{a^{-1}ma \mid a \in G\}$ is called the conjugate class of m. More generally, if H is any subgroup of G, the set $W_H(m) = \{h^{-1}mh \mid h \in H\}$ is called a Wigner H-class. The set $\{e\}$ is always a Wigner H-class and, if G is abelian then $W_H(m) = \{m\}$ for each $m \in G$.

4.2   Let H be a subgroup of the finite group G. Each $m \in G$ belongs to at least one Wigner H-class (for example, $W_H(m)$) and any two Wigner H-classes are either identical or disjoint. Therefore G is the pairwise disjoint union of all the distinct Wigner H-classes.

Distinct Wigner H-classes can contain different numbers of elements; in fact,

$$4.3 \qquad |W_H(m)| = [H:H \cap C_G(m)].$$

It is easy to verify that a subgroup $K \subset H$ is normal in H if and only if K is a (pairwise disjoint) union of Wigner H-classes.

5. Homomorphism

Let $(G,\cdot)$ and $(L,*)$ be any two groups. A mapping $f:G \to L$ is called a homomorphism if $f(a\cdot b) = f(a)*f(b)$ for all $a,b \in G$. We say that a homomorphism $f:G \to L$ is monic (or: a monomorphism) if the map f is injective; it is called epic (or: an epimorphism) whenever the map f is surjective. The homomorphism $f:G \to L$ is called an isomorphism whenever f is bijective; in this case we write $f:G \cong L$ and call the groups isomorphic. If $f:G \to L$ is an isomorphism, then the inverse map $f^{-1}:L \to G$ is also an isomorphism.

Whenever two groups G and L are isomorphic, any equality in $(G,\cdot)$ involving only the elements and the group operation can be transferred, via an isomorphism f, to an analogous equality in $(L,*)$; and conversely. Thus, there is no algebraic distinction between isomorphic groups; their only difference is in the names given to the elements, and the way in which the group operation is denoted.

If two groups are isomorphic, they must have the same order; but the converse is not necessarily true. However, for each integer n, there is only a finite number $\gamma(n)$ of non-isomorphic groups of order n. Although $\gamma(n)$ is known for many n (for example, $\gamma(4) = 2$, $\gamma(8) = \gamma(12) = 5$,

$\gamma(16) = 14$) little is known about $\gamma(n)$ in general, other than that $\gamma(n) = 1$ if and only if n is of the form $n = p_1 \bullet \dots \bullet p_s$, where the primes p_i are all different and no p_i-1 is divisible by any p_j (so, for example, $\gamma(n) = 1$ whenever n is a prime, or if n = 15, or n = 455).

The following results are standard

5.1   Let f:G → L be any homomorphism. Then for each subgroup H ⊂ G the image f(H) is a subgroup of L.

5.2   Let N be a normal subgroup of G, and let G/N be the factor group of G by N. Then the map p:G → G/N defined by a ↦ (Na) is an epimorphism.

5.3   Let N ◁ G and let K be any subgroup of G. Then K·N = ⟨K,N⟩ and K·N/N ≅ K/K ∩ N

5.4   Let N ◁ G, and let H ◁ G be such that N ⊂ H. Then

$$ G/_H \quad \cong \quad \frac{G/_N}{H/_N} . $$

An isomorphism f:G → G of a group G onto itself is called an automorphism of G. For any group G, the identity map id:G → G defined by x ↦ x is an automorphism; for any element a ∈ G, the map x ↦ $a^{-1}xa$ of G onto itself is also an automorphism, called an inner automorphism. Any automorphism not an inner automorphism is called an outer automorphism; if G is abelian then every automorphism other than the identity automorphism, is an outer automorphism.

<u>6.</u>   <u>Direct and semidirect Products</u>   In this section, we seek to express the elements of a group as products of elements taken from appropriate subgroups; this frequently simplifies the construction and handling of large groups, and serves to clarify their structure.

(a)   We say that a group G is the direct product of a normal subgroup $N \triangleleft G$ and a normal subgroup $K \triangleleft G$ (and write $G = N \times K$) whenever both $G = N \cdot K$ and $N \cap K = \{e\}$.

In case $G = N \times K$, each $g \in G$ is uniquely representable as a product $g = n \cdot k$ where $n \in N$ and $k \in K$. Moreover each product $nk = kn$: for, $n^{-1}k^{-1}nk = (n^{-1}k^{-1}n) \cdot k \in K$ because $K \triangleleft G$ and $n^{-1}k^{-1}nk = n^{-1}(k^{-1}nk) \in N$ because $N \triangleleft G$, so $n^{-1}k^{-1}nk = \{e\}$ and $nk = kn$. Finally, $G/N \cong K$. Using the unique representation of the elements of G as products $n \cdot k$, we can express the group operation in G directly in terms of these products: if $g_0 = n_0 k_0$ and $g_1 = n_1 k_1$, then $g_0 g_1 = (n_0 n_1) \cdot (k_0 k_1)$.

We now relax the condition that both subgroups are normal

(b)   We say a group G is the semidirect product of a normal subgroup $N \triangleleft G$ and a subgroup $H \subset G$ (and write $G = N \wedge H$) whenever both $G = N \cdot H$ and $N \cap H = \{e\}$.

In case $G = N \wedge H$, each $g \in G$ is uniquely representable as a product $g = n \cdot h$ where $n \in N$ and $h \in H$, and g is also uniquely representable as a product $g = h' \cdot n'$; however, unlike the case in direct products, it is not necessarily true that $h = h'$ and $n = n'$; in fact, if $g = nh = h'n'$, then $h' = h$ and $n' = h^{-1}nh$ (which is an element of N because $N \triangleleft G$). Using the unique representation of the elements of G as products $n \cdot h$, we can express the group operation directly in terms of these products: if $g_0 = n_0 h_0$ and $g_1 = n_1 h_1$, then $g_0 g_1 = (n_0 \cdot [h_0 n_1 h_0^{-1}]) \cdot (h_0 h_1)$.

To give some conditions under which a given group G has a direct and/or semidirect product decomposition, it is convenient to start with a notion not involving normality:

(c) Let G be a group and $H \subset G$ a subgroup. We say that H is complemented

in G if there is some subgroup $K \subset G$ such that both $G = H \cdot K$ and

$H \cap K = \{e\}$.

The subgroup K is called a complement of H; it is not in general uniquely determined by H. In fact, if K is a complement of H, so also is each conjugate $g^{-1} K g$; moreover, a given H may have complements K, K' that are not isomorphic, or it may not have any complement.

The following is a simple characterization of those subgroups of G that have complements; it is essentially a restatement of (c).

6.1   Let G be a finite group and $H \subset G$ a subgroup. Then H has a complement in G if and only if H has a right transversal T in G which is a subgroup of G.

Proof   If H has a right transversal T that is a group, then (2.2 et seq.) $G = H \cdot T$ and $H \cap T = \{e\}$, so T is a complement. Conversely, suppose that $G = H \cdot K$ and $H \cap K = \{e\}$; we show that K is a right transversal of H in G.
(a) Any coset $Ha = Hk$ for a suitable $k \in K$: we need only write $a = hk$.
(b) No coset Ha contains more than one $k \in K$: if $k_0, k_1 \in Ha$, then $k_0 k_1^{-1} \in H$

and, since $k_0 k_1^{-1} \in K$, we get $k_0 k_1^{-1} = e$ so $k_0 = k_1$. Thus, the group K is in fact a transversal of H in G.

Applying this result in the case $H \lhd G$, and recalling 5.3, leads to a result we use frequently

6.2   Let G be a finite group and let $N \lhd G$. Then G has a semidirect product representation $G = N \wedge K$ if and only if N has a transversal K that is a group. Moreover, every complement K of N is isomorphic to the group G/N.

If G is known to be the direct product $N \times K$, then a group $\hat{G}$ isomorphic to G can be constructed directly from the groups N and K alone. The elements of $\hat{G}$ are all ordered pairs (n,k), where $n \in N$ and $k \in K$, and the multiplication in $\hat{G}$ is defined by $(n_0, k_0)(n_1, k_1) = (n_0 n_1, k_0 k_1)$. The map $g = nk \mapsto (n,k)$ shows G isomorphic to $\hat{G}$ with N corresponding to the subgroup $\{(n,e) \mid n \in N\} \subset \hat{G}$ and K to the subgroup $\{(e,k) \mid k \in K\} \subset \hat{G}$.

If G is known to be the semidirect product $N \wedge H$, then again a group $\hat{G}$ isomorphic to G can be constructed using only N and H. The elements of $\hat{G}$ are all ordered pairs (n,h) where $n \in N$ and $h \in H$, and this time the multiplication in $\hat{G}$ is defined by $(n,h) \cdot (r,t) = (n \cdot h r h^{-1}, ht)$. The map $g = nh \mapsto (n,h)$ shows G isomorphic to $\hat{G}$, with N corresponding to the normal subgroup $\{(n,e) \mid n \in N\} \subset N \wedge H$ and H to the subgroup $\{(e,h) \mid h \in H\} \subset \hat{G}$.

7.   **Permutations**   Let L be any set. By a permutation λ of L is meant a bijection $\lambda : L \to L$. If μ is another permutation of L, then the composition $\lambda \circ \mu : L \to L$ is also a permutation of L. With the binary operation

213

$\lambda \circ \mu = \lambda \cdot \mu$, the set of all permutations of L forms a group Sym(L), called

the symmetric group of L; the identity of Sym(L) is the identity map

id:L → L, and the inverse of any given $\lambda \in$ SymL is the inverse λ^{-1} of the

bijection λ.

Whenever L is a finite set, a permutation $\lambda: L \to L$ can be conven-

iently expressed by a 2 x n matrix

$$\lambda = \begin{pmatrix} 1_1, \ldots, 1_s \\ \lambda(1_1), \ldots, \lambda(1_s) \end{pmatrix}$$

where the elements of L are listed in the first row, and the image of each

$1 \in$ L under λ is written directly below it. We note that if $|L| = n$, then

$|\text{SymL}| = n!$, and that SymL is not abelian whenever $|L| \geq 3$.

If $L = \{1, 2, \ldots, n\}$, we frequently denote SymL by S_n and call it the

symmetric group on n letters. The subset of all permutations in S_n that

leave the letter n fixed is a subgroup of S_n isomorphic to S_{n-1}.

Special types of permutations, called cycles, play an important rôle.

A permutation $\lambda \in$ SymL that permutes a subset $\{1_1, 1_2, \ldots, 1_{s-1}, 1_s\} \subset L$

cyclically, in the sense that $\lambda(1_1) = 1_2, \lambda(1_2) = 1_3, \ldots, \lambda(1_{s-1}) = 1_s, \lambda(1_s)$

$= 1_1$, and leaves every other element of L fixed is called a cycle, and the

number s of elements in the cycle is called its length. Clearly, any cycle

of length 1 is the identity map.

A cycle λ can be written simply as $(1_1, 1_2, \ldots, 1_s)$ to indicate that

$1_1 \mapsto 1_2, 1_2 \mapsto 1_3, \ldots, 1_s \mapsto 1_1$; the same cycle can be written in many

ways, for example $(1_1, 1_2, \ldots, 1_s) = (1_2, \ldots, 1_s, 1_1)$ since it is the circular

order of the elements in the cycle that determines the permutation. If

$\lambda = (s_1, s_2, \ldots, s_a)$ and $\mu = (t_1 t_2, \ldots, t_b)$ are two cycles, the product $\mu \cdot \lambda$ is written $(t_1, t_2, \ldots, t_b) \cdot (s_1, s_2, \ldots, s_a)$, so that the products of cycles correspond to their consecutive action, proceeding from left to right, i. e. perform the permutations in the order that they are written[*].

The powers of a cycle are easily calculated: if $\lambda = (l_1, \ldots, l_s)$ then λ^p maps each l_i to the pth element following it in the circular ordering of $(l_1, \ldots, l_s)$; in particular, the order of any cycle is its length, and the inverse of any cycle is that cycle written in the reverse circular order.

Two or more cycles with no elements in common, are called disjoint; if λ, μ are disjoint cycles, then $\lambda\mu = \mu\lambda$, since each permutation changes only elements left fixed by the other.

Given any permutation $\lambda: L \to L$, we can choose any $l \in L$, then write down its image, and next the image of that, proceeding until the cycle closes; choosing an element of L not used and repeating the process, we are led to

7.1 Every $\lambda \in \mathrm{Sym}\, L$ not the identity permutation can be expressed uniquely, exept for order, as a composition of disjoint cycles, each of length ≥ 2.

Thus, for example, $\begin{pmatrix} 12345 \\ 21435 \end{pmatrix} = (12)(34)$.

(*) For example, the action $(123)(124)$ on E results from the action of (124) on E which yields $(124)E$, followed by the action of (123) on $(124)E$, leading ultimately to $(14)(23)E$. Note that the permutation $(14)(23)$ is found through the mappings $\begin{pmatrix} 12345 \\ 23145 \\ 43215 \end{pmatrix}$.

This corresponds to first permuting the ordered set $\{1,2,3,4,5\}$ according to (123), and then (124).

This representation of a permutation λ as the composition of disjoint cycles is called the cycle decomposition of λ, and is particularely convenient.

7.2 The order of a permutation λ is the least common multiple of the lengths of the cycles in its cycle decomposition.

If a permutation s has the cycle decomposition $s = (s_1,\ldots,s_k)\ldots(t_1,\ldots,t_r)$ then the conjugate $\lambda^{-1}s\lambda = (\lambda(s_1),\ldots,\lambda(s_k))\ldots(\lambda(t_1),\ldots,\lambda(t_r))$: it is obtained by replacing each member in each cycle by its image under λ; geometrically, this represents the motion s applied on λ. In particular, whenever two permutations have the same cycle structure, their only difference is the labeling, and since they can be made identical by a permutation, it follows that

7.3 Two elements λ,μ ∈ SymL belong to the same conjugate class if and only if they have the same cycle structure.

A cycle of length 2 is called a transposition. Each permutation λ:L → L can be written as a product of transpositions (which are not in general disjoint). This may be done in many ways, but for any given λ the number of transpositions required will always be even, or always odd. A permutation is called even (resp. odd) if the number of transpositions is even (resp. odd). Any transposition is an odd permutation; and since any cycle (12...k) of length k can be written (12...k) = (12)(13)...(1k) as a product of (k-1) transpositions, a cycle of length k is even or odd

according as k-1 is even or odd. From this and the cycle decomposition of a

$\lambda \in \text{Sym} L$ follows

7.4   If λ is written as a product of disjoint cycles of lengths $l_1,\ldots,l_s$,

then λ is even or odd according as $N(\lambda) = \sum_{i=1}^{s} (l_i-1)$ is even or odd.

The set of all even permutations in S_n form a subgroup, called the

alternating group on n letters and denoted by A_n. The alternating group A_n

is of index 2 in S_n, therefore it is a normal subgroup of S_n; its coset is

the set of all odd permutations. The group A_n is simple for all $n \geq 5$.

We know (2.3) that if H is a subgroup of a finite group G, then $|H|$

divides $|G|$. The converse is not in general true: if n is an integer that

divides $|G|$, then G may not have a subgroup of order n. The standard

example is A_4: here, $|A_4| = 12$, and A_4 contains subgroups of order 2 and 3,

but A_4 has no subgroup of order 6. We use this fact to prove a result

needed for our discussion of the asymmetric carbon atom.

7.5   The only subgroup of order 12 in S_4 is A_4.

Proof   Assume there was such a subgroup N not contained in A_4. Since A_4

is normal, $N \cdot A_4$ is a subgroup of S_4, and since it contains A_4 properly, it

must have order 24, so $N \cdot A_4 = S_4$. Let $n = |N \cap A_4|$; then

$$24 = |N \cdot A_4| = \frac{|N| \, |A_4|}{|N \cap A_4|} = \frac{12 \cdot 12}{n}$$

showing $n = 6$. Thus, $N \cap A_4$ would be a subgroup of order 6 in A_4, and this is impossible. Therefore $N \subset A_4$ and, being of order 12, we have $N = A_4$.

8. **Burnside-Frobenius Theorem**   In this section we consider the Burnside-Frobenius theorem, and two consequences, the enumeration of Wigner classes and the enumeration of double cosets. Since a portion of this material is not readily available in the standard references, some proofs will be given.

Let X be any set and G any group. By an action of G on X is meant a rule that assigns to each $g \in G$ a map $T_g : X \to X$ such that

   (i) $T_{gh} = T_g \cdot T_h$ for every $g, h \in G$

and (ii) $T_e(x) = x$ for all $x \in X$.

An action of G on X is called trivial if each T_g is the identity map. The action is called transitive if for each pair $x, y \in X$, there is some $g \in G$ with $T_g(x) = y$; and the action is called effective if $T_g = T_h$ implies that $g = h$.

Let G act on X. Since $T_{g^{-1}} \cdot T_g = T_{g^{-1}g} = T_e = $ id and $T_g \cdot T_{g^{-1}} = $ id, it follows that each $T_g : X \to X$ is bijective, and that $T_g^{-1} = T_{g^{-1}}$. In the sequel, we shall denote the map $T_g : X \to X$ simply by $g : X \to X$, and the element $T_g(x)$ by gx.

Let G act on X. For each $x \in X$, the set $Gx = \{gx \mid g \in G\}$ is called the orbit of x under G. Note that $Gy = Gx$ for each $y \in Gx$: for, if $y = gx$, then $Gy = Ggx = (Gg)x = Gx$ because G is a group.

8.1   Let G act on X. Each x ε X belongs to exactly one orbit, namely Gx, and any two orbits are either identical or disjoint. Therefore X is the pairwise disjoint union of all distinct orbits. Different orbits may contain different numbers of elements.

To determine the size of an orbit, we need an additional concept. For each x ε X,   the set $\text{Stab}(x) = \{g ε G \mid gx = x\}$ is called the stabilizer of x.

8.2   Stab(x) is a subgroup of G. For any x ε X and g ε G, Stab(gx) = $g[\text{Stab}(x)]g^{-1}$ and, in particular, $|\text{Stab}(x)| = |\text{Stab}(gx)|$ for all x ε X and g ε G.

We can now express the size of the orbit Gx in terms of Stabx:

8.3   Let a finite group G act on a finite set X. Then $|Gx| = \dfrac{|G|}{|\text{Stab}(x)|}$ for each x ε X.

<u>Proof</u> Let x be given. We shall show that gx = hx if and only if g and h belong to the same left coset of Stabx in G: For, if gx = hx, then $h^{-1}gx = x$ so $h^{-1}g ε$ Stabx and h,g are the same left coset of Stabx; the converse follows by reversing the steps of this argument. With this we have shown that there are as many distinct elements in the orbit Gx as there are left cosets of Stabx in G so, by 2.3, we find |Gx| = |G|/|Stabx| and the proof is complete.

The Burnside-Frobenius theorem, which we will now prove, is very useful in questions of enumeration. In order to formulate it, we need the following notion: for each $g \in G$, the set $\text{Fix}(g) = \{x \in X \mid gx = x\}$ is called the fixed-point set of g.

8.4 (Burnside-Frobenius) Let G be a finite group acting on a finite set X. Then, the total number T of distinct orbits is

$$T = \frac{1}{|G|} \sum_{g \in G} |\text{Fix}(g)| = \frac{1}{|G|} \sum_{x \in X} |\text{Stab}(x)|$$

<u>Proof</u> Consider the array

	e	g_1	g_2	.	.
x_1	'				
x_2	'	'			
x_3	'				
.					
.					

and place a check in row x and column g whenever gx = x. We count the number of checks in three different ways. First: counting by columns we have $\sum_{g \in G} |\text{Fix}(g)|$ checks in the array. Second: counting by rows, we have $\sum_{x \in X} |\text{Stab}(x)|$ checks in the array, so these two sums are equal. Finally, we count the checks by considering orbits. Fix attention on row x; a check will appear whenever gx = x so, in that row, we will find $|\text{Stab} x|$ checks. By 8.2, each $y \in Gx$ will also have $|\text{Stab}(x)|$ marks so each orbit Gx will contribute $|Gx| \cdot |\text{Stab} x|$ marks which, by 8.3, means that each orbit will contribute $|G|$ marks. Thus, if there are T orbits, the number of checks in the table must be $T \cdot |G|$ and comparing this count with the previous counts, the proof is complete.

We give two applications of this result

8.5 (The number of Wigner H-classes) Let G be a finite group and H a sub-group. The total number T of distinct Wigner H-classes is

$$T = \frac{1}{|H|} \sum_{h \in H} |C_G(h)|$$

Proof For each $h \in H$, let $T_h : G \to G$ be the map $x \mapsto h^{-1}xh$; it is easy to verify that this determines an action of H on G. With this action, the orbit Hx of any $x \in G$ is precisely the Wigner H-class $W_H(x)$. Now $Fix(h) = \{x \in G \mid h^{-1}xh = x\} = \{x \in G \mid xh = hx\} = C_G(h)$, the centralizer of h in G and so, by the Burnside-Frobenius theorem, the result follows.

8.6 (The number of double cosets) Let G be a finite group and H,K two subgroups. The number T of distinct (H,K)-double cosets is

$$T = \frac{1}{|H|} \sum_{i=1}^{n} |Hg_i \cap g_iK|$$

where $\{g_1, \dots, g_n\}$ is a left transversal of K in G.

Proof Let X be the set of distinct left cosets of K, and for each $h \in H$ let $T_h : X \to X$ be the map $[gK] \mapsto [hgK]$. This map is well-defined, i. e. $T_h[gK]$ depends on the coset $[gK]$ and not on the representative g used for $[gK]$: indeed, if $gK = \hat{g}K$, then $g^{-1}\hat{g} \in K$ so $g^{-1}h^{-1}h\hat{g} \in K$ and therefore $[h\hat{g}K] = hgK]$.

It is easy to verify that the correspondence $h : T_h$ is an ac,tion of H on X. We observe that the orbit of $[gK]$ is $H[gK] = \{[hgK] \mid h \in H\} = HgK$, i. e. the orbit of gK is precisely the set of all cosets $\hat{g}K$ making up the

221

double coset HgK.   To   find   the   number of distinct orbits by using the

Burnside-Frobenius theorem,   we need only find the stabilizer of each   left

coset $[gK]$. For this purpose,   let $g_1,\ldots,g_n$ be a left transversal of K

in G.   Then $\text{Stab}[g_iK] = \{h|hg_iK = g_iK\} = \{h|h \in g_iKg_i^{-1}\}$.   This   says   that

$h \in \text{Stab}[g_iK]$ if and only if $h \in g_iKg_i^{-1} \cap H$ so $\text{Stab}[g_iK] = H \cap g_iKg_i^{-1}$ and,

since $|H \cap g_iKg_i^{-1}| = |Hg_i \cap g_iK|$ an application of the   Burnside-Frobenius

theorem completes the proof.

9. Generalized wreath products.   We collect here the general results of a

group construction used in the text;   since this material has not appeared

elsewhere in this form, proofs are included.

Let L be a finite set and $\{L_1,\ldots,L_n\}$ a partition of L, so that L is

the pairwise disjoint union $L = \bigcup_1^n L_i$. The $|L_i|$ need not have any common

value.

Let $A \subseteq \text{Sym}L$ be the set of all permutation h with the property that

$h(L_i) = L_i$ for each $i = 1,\ldots,n$; in symbols,

$$A = \{h \in \text{Sym}L \mid h(L_i) = L_i \quad \text{for each } i=1,\ldots,n\}$$

This is a subgroup of $\text{Sym}L$, since the composition of any two such

permutations has the same property.   In fact, A is isomorphic to the

direct product $\prod_1^n \text{Sym}L_i$.

Let $N \subseteq \text{Sym}L$ be the set of all permutations that map each L_i onto some

L_j, i. e.

$$N = \{\alpha \in \text{Sym}L \mid \text{for each } i \text{ there is some } j \text{ with } \alpha(L_i) = L_j\}$$

Because each $\alpha \in \text{Sym} L$ is bijective, this condition implies, among other things, that for each L_i, the map $\alpha | L_i$ is a permutation sending L_i onto some L_j having $|L_j| = |L_i|$, and that no two distinct L_i are mapped into a single L_j. It is immediate that N is a subgroup of $\text{Sym} L$, because the composition of any two $\alpha, \beta \in N$ is also a member of N. The group N is a generalized wreath product.

9.1 A is a normal subgroup of N.

Proof It is clear from the definitions that A is a subgroup of N. To show it normal, let $h \in A$ and $\alpha \in N$ be given. Fixing attention on any L_i, we have an L_j with $\alpha(L_i) = L_j$, so that $h\,\alpha(L_i) = h(L_j) = L_j$ and therefore $\alpha^{-1}h\alpha(L_i) = \alpha^{-1}(L_j) = L_i$. This holds for each L_i, so $\alpha^{-1}h\alpha \in A$ and therefore $A \triangleleft N$.

Suppose $\alpha \notin N$; then there is some L_i such $\alpha(L_i) \neq L_j$ for all j. This can occur only if $\alpha(L_i)$ meets more than one L_j or if $\alpha(L_i)$ is a proper subset of some L_j; in the latter case, $\alpha^{-1}(L_j)$ will meet more than one L_i. Thus, if $\alpha \notin N$, there is some L_i such that either $\alpha(L_i)$ or $\alpha^{-1}(L_i)$ meets more than one set L_j. We use this observation to show

9.2 N is the normalizer of A in $\text{Sym} L$.

Proof Since $A \triangleleft N$, it is enough to show that, if $\alpha \notin N$, then $\alpha h \alpha^{-1} \notin A$ for some $h \in A$. Using our observation, we can assume that there is some L_i such that $\alpha(L_i)$ meets two distinct sets L_j and L_s (otherwise we work with

α^{-1}). Choose $v, w \in L_i$ so that $x = \alpha(v) \in L_j$ and $\alpha(w) \in L_s$. Now let $h \in A$ be the permutation that interchanges v, w, keeping all other elements of L fixed. Then $x \in L_j$ and since $h\alpha^{-1}(x) = h(v) = w$ we have $\alpha h\alpha^{-1}(x) = h(w) \in L_s$. Thus, $\alpha h \alpha^{-1}$ does not map L_j into itself, so $\alpha h\alpha^{-1} \notin A$ and α is not in the normalizer of A. If we worked with α^{-1}, the argument just given shows α^{-1} is not in the normalizer of A; since the normalizer is a group, it follows that α cannot belong to the normalizer.

9.3 If each $|L_i| \geq 2$, then N is its own normalizer in SymL.

Proof Let $\alpha \notin N$. With the notation in the proof of 9.2 we shall show that, if $|L_j| \geq 2$, then in fact $\alpha h\alpha^{-1} \notin N$; since $h \in A \subseteq N$, this will establish that α is not in the normalizer of N. We have found an $x \in L_j$ such that $\alpha h\alpha^{-1}(x) \in L_s$. Since $|L_j| \geq 2$, there is a $y \in L_j$ with $y \neq x$; since $\alpha^{-1}(y) \neq v$ or w, we have $h\alpha^{-1}(y) = \alpha^{-1}(y)$ so that $\alpha h\alpha^{-1}(y) = y$. Thus, $\alpha h\alpha^{-1}$ does not send L_j into a single set L_s, so $\alpha h\alpha^{-1} \notin N$.

Because $A \triangleleft N$, the distinct cosets of A in N, with the usual coset multiplication, form a group $N/_A$. To characterize $N/_A$ we need a direct description of the cosets:

9.4 $\alpha, \beta \in N$ belong to the same coset of A if and only if $\alpha(L_i) = \beta(L_i)$ for each $i = 1, \dots, n$.

Proof Let $\alpha, \beta \in N$ belong to the same coset of A (we can use right or left cosets because $A \triangleleft N$). Then $\beta = \alpha h$ for some $h \in A$, so for each $i =$

$1,\ldots,n$ we have $\beta(L_i) = \alpha\, h(L_i) = \alpha(L_i)$. Conversely, if $\beta(L_i) = \alpha(L_i)$ for

each i then $\alpha^{-1}\beta(L_i) = L_i$ for each i so $\alpha^{-1}\beta = h \in A$ and therefore α, β

belong to the same coset of A.

Now consider the set $B = \{1,\ldots,n\}$ of indices of the sets L_i. Break

up $\{1,\ldots,n\}$ into blocks $B_1,\ldots,B_k$ by placing in the same block all the

indices i for which the $|L_i|$ have the same value; the $B_1,\ldots,B_k$ are a

partition of B, and there are as many different B_i as there are different

values of $|L_i|$. Consider now the set of all permutations $Q' = \{g \in \mathrm{Sym}\,B \mid$

$g(B_i) = B_i$ for each $i = 1,\ldots,k\}$. This is clearly a subgroup of $\mathrm{Sym}\,B$, and

we have

9.5 $Q' \cong N/A$

Proof   For each $\alpha \in N$, let $\pi\alpha \in Q'$ be the permutation defined by setting

$\pi\alpha(i) = j$ if $\alpha(L_i) = L_j$, i.e. $\pi\alpha$ is the permutation of the sets L_i

associated with $\alpha \in N$. It is easy to see that $\pi(\alpha\cdot\beta) = \pi\alpha\cdot\pi\beta$, so $\pi: N \to Q'$

is a homomorphism. By 9.4, $\pi\alpha$ is the same for all α belonging to a common

coset of A, so π induces a homomorphism $\hat{\pi}$ of the group N/A into Q'. It is

obvious, again from 9.4, that $\hat{\pi}$ is bijective so $\hat{\pi}$ is an isomorphism.

9.6   Let $L = \overset{n}{\underset{1}{\bigcup}} L_i$ and let $c_1,\ldots,c_k$ be the number of distinct cardinals

$|L_i|$. For each c_i assume that there are b_i sets L_i having c_i elements

(i.e. cardinal c_i). Then $|N| = |L_1|!\ldots|L_n|!\ b_1!\ldots b_k!$

Proof   We have $|N/A| = |Q'| = b_1!\ldots b_k!$; since $|A| = |L_1|!\ldots|L_n|!$ and

$|N| = |^N/_A| \cdot |A|$, the proof is complete.

We finally establish

9.7   N is a semidirect product $N = A \wedge Q$, where $Q \cong Q' \cong {}^N/_A$.

Proof   Our objective (cf 6.2) is to pick a transversal of A in N that is a group, and we begin by linearly ordering the elements of L. For each permutation $q \in Q'$, pick $\sigma(q) \in N$ as follows: for each i, $\sigma(q)$ is the unique order-preserving map of L_i onto $L_{q(i)}$. It is immediate from 9.4 and the definition of Q' that $\{\sigma(q) \mid q \in Q\}$ is a transversal of A in N. It is straightforward to verify that $\sigma(q \cdot q') = \sigma(q) \cdot \sigma(q')$ so the map $\sigma: Q' \to N$ is a homomorphism, therefore by 5.1, the transversal $\{\sigma(q) \mid q \in Q'\}$ is indeed a subgroup of N. This completes the proof.

10. Primitive and imprimitive actions   Let G be a group acting on a set X. By a block partition $\{B_1, \ldots, B_n\}$ of X relative to G is meant a decomposition of X into pairwise disjoint sets, $X = \overset{n}{\underset{1}{\cup}} B_i$, such that every $g \in G$ maps each B_i either onto itself or onto some other B_j; each $g \in G$ therefore permutes the blocks B_i as wholes.

There are always two trivial block partitions of X: that in which the blocks are all 1-element sets, and that having the single block X; block partitions are therefore not unique. If there are no non-trivial block partitions of X relative to G, the group G is said to act primitively on X; otherwise the action on G is called imprimitive.

Given an action of G on X, define a relation $\sim$ in X by $a \sim b$ if there is some $g \in G$ with $g(a) = b$. This is an equivalence relation in X; the equivalence classes are called the orbits (or: transitivity domains) of G, and for each transitivity domain T we have $g(T) = T$ for every $g \in G$. Thus, if G does not act transitively on X, the transitivity domains form a non-trivial block partition of X, and the action of G is therefore imprimitive.

However, even though G acts transitively on a set X, it can still act imprimitively. For example, the action on $X = \{1,2,3,4\}$ by the cyclic group $\mathbb{Z}_4$ generated by the permutation (1234) is transitive, but it is imprimitive with X having the block partition $\{1,3\} \cup \{2,4\}$. In particular, although G acts transitively on each transitivity domain T, it can still act imprimitively on T.

We first consider how imprimitivity can arise for transitive actions; for our purpose, the following special result is needed:

10.1 <u>Theorem</u> Let G act transitively on X, choose an $a \in X$, and let $\Sigma_a = \{g \in G \mid g(a) = a\}$ be the stabilizer of a. Let H be any maximal subgroup of G containing Σ_a and define $\Delta = \{h(a) \mid h \in H\}$. If T is a left transversal of H in G, then $\{t\Delta \mid t \in T\}$ forms a block partition of X relative to G. This consists of $[G:H]$ blocks, each containing $[H:\Sigma_a]$ elements. Moreover, for each block $t\Delta$ we have $\{g \in G \mid g(t\Delta\} = tHt^{-1}$; in particular, H is the stabilizer of the set Δ.

<u>Proof</u> We first show $X = \cup \{t\Delta \mid t \in T\}$. For, given $b \in X$, there is, by transitivity, some $g \in G$ with $g(a) = b$; this g belongs to some coset tH, so $g = t \cdot h$ for some $h \in H$, therefore $b = g(a) = t \cdot h(a) \in t\Delta$.

We next show that the family $\{t\Delta \mid t \in T\}$ is pairwise disjoint. Assume $b \in t\Delta \cap t_0\Delta$; then $b = t \cdot h(a) = t_0 \cdot h_0(a)$ for suitable $h, h_0 \in H$. This shows $h_0^{-1} t_0^{-1} t \cdot h(a) = a$, so that $h_0^{-1} t_0^{-1} t h \in \Sigma_a \subset H$, therefore $t_0^{-1} t \in H$ and $t \in t_0 H$. Since T is a left transversal, this means that $t = t_0$, so $t\Delta = t_0\Delta$.

The family $\{t\Delta \mid t \in T\}$ is a block partition of X relative to G. For, let $g \in G$ and assume that $g(t\Delta) \cap t_0\Delta \neq 0$; then there is some $b = g \cdot t \cdot h(a) = t_0 h_0(a)$ which implies, as above, that $h_0^{-1} t_0^{-1} g t h \in \Sigma_a \subset H$ and therefore that $g \cdot t \in t_0 H$, say $g \cdot t = t_0 h_1$. Then $g(t\Delta) = t_0 h_1(\Delta) = t_0 h_1 \{h(a) \mid h \in H\} = t_0\Delta$. Thus, each $g \in G$ maps each block $t\Delta$ either onto itself or onto some other block $t_0\Delta$.

If $g(t\Delta) = t\Delta$ the same argument shows $g \in tHt^{-1}$; conversely, if $g \in tHt^{-1}$, then $gt \in tH$, so $g \cdot t = th_1$ for some $h_1 \in H$ and $g(t\Delta) = t\Delta$. Thus, tHt^{-1} is the stabilizer of the block Δ.

Each block contains $[H:\Sigma_a]$ elements: for $h(a) = h_0(a)$ if and only if $h_0^{-1} h(a) = a$, that is, $h_0^{-1} h \in \Sigma_a$ or, equivalently, h, h_0 belong to the same left coset of Σ_a in H. That there are $[G:H]$ distinct blocks now follows immediately.

The block partition in the theorem depends on the maximal subgroup $H \supset \Sigma_a$ that is selected; distinct such maximal subgroups give different block partitions. Observe that if Σ_a is itself a maximal subgroup of G, then $\{h(a) \mid h \in \Sigma_a\} = a$, and the resulting block partition is the trivial one consisting of singletons. It is fairly easy to see that the action of G on X is primitive if and only if the stabilizer Σ_a of any (and therefore of each) $a \in X$ is a maximal subgroup of G.

Returning to the theorem, we now consider how G permutes the blocks.

Let $\hat{X}$ be the set $\{t^\Delta \mid t \in T\}$ of blocks. Defining $g\{t^\Delta\} = \{g(t^\Delta)\}$ determines an action of G on $\hat{X}$ and we have

10.2 Corollary The group G acts transitively and primitively on $\hat{X}$. The set of all $g \in G$ that induce the identity map of $\hat{X}$ is precisely the core of H in G, so is a normal subgroup of G.

Proof The transitivity of G on $\hat{X}$ is immediate from that of G on X. Assume that the action on $\hat{X}$ is not primitive. Then $\hat{X}$ has a block partition $\{B_1,\ldots,B_n\}$ relative to G; let $\Delta \in B_1$. It is clear that $\{g \in G \mid gB_1 = B_1\}$ is a group $\hat{H} \subseteq G$ and, in fact, a proper subgroup because G is transitive on $\hat{X}$. Now, if $h \in H = \text{stab}^\Delta$, we have $h(\Delta) = \Delta$ so, in the action on $\hat{X}$ we find $h(B_1) \cap B_1 \neq \emptyset$ therefore $h(B_1) = B_1$. Thus, $H \subseteq \hat{H}$; moreover, since B_1 is a non-trivial block and G is transitive on X, we find that H is a proper subgroup of $\hat{H}$. This contradicts the maximality of H, and proves that G acts primitively on X. The last part of the theorem follows by noting that $\{g \in G \mid g(t^\Delta) = t^\Delta \text{ for each } t \in T\} = \{tHt^{-1} \mid t \in T\}$ and using 3.7.

With these results, we can now describe the nature of any action of a finite group on a set X. First, find the transitivity domains $T_1,\ldots,T_s$ of the action of G on X; then G acts transitively on each set T_i by $g \mapsto g \mid T_i$. For each $i = 1,\ldots,s$, choose a maximal subgroup H_i containing the stabilizer of some element $a_i \in T_i$; this gives a block partition of T_i into $\Delta_{i1},\ldots,\Delta_{in_i}$. It follows that every $g \in G$ maps each block onto a block; and $F = \overset{s}{\underset{1}{\bigcap}} \operatorname{core}_G(H_i)$ is precisely the set of all $g \in G$ such that $g(\Delta_{ij}) = \Delta_{ij}$ for each block Δ_{ij}; being the intersection of normal subgroups, F is itself a normal subgroup of G. We summarize this discussion in a form needed for the text:

10.3 <u>Theorem</u> Let the finite group G act on X then

1. The action determines a uniquely defined partition $X = \overset{s}{\underset{1}{\bigcup}} T_i$ of X, where the T_i are the transitivity domains of the action

2. Each T_i has a (perhaps trivial) block partition, $T_i = \overset{n_i}{\underset{k=1}{\bigcup}} \Delta_{ik}$ determined by a maximal subgroup H_i containing the stabilizer of some $a_i \in T_i$

3. Every $g \in G$ maps the blocks in each T_i onto blocks in T_i

4. The subgroup $F \subseteq G$ mapping every block Δ_{ij} onto itself is a normal subgroup of G, and is determined by the maximal subgroups H_i that were selected in step 2.

References

A. Ian D. Macdonald, The Theory of Groups, Clarendon Press)xford, 1968.

B. John S. Rose, A course on Group Theory, Cambridge University Press 1978.

C. B. Huppert, Endliche Gruppen I, Springer-Verlag, Heidelberg 1967.

D. S. MacLane & G. Birkhoff, Algebra, Macmillan, New York 1967.

AUTHOR INDEX

SUBJECT INDEX

ERRATUM

Lecture Notes in Chemistry, Vol. 36
I. Ugi
J. Dugundji
R. Kopp
D. Marquarding
Perspectives in Theoretical Stereochemisty
ISBN 3-540-13391-7

The second author on the list must read
in the prelims as follows:

J. D U G U N D J I

© Springer-Verlag Berlin Heidelberg 1984

Lecture Notes in Chemistry

Editors: G. Berthier, M. J. S. Dewar, H. Fischer, K. Fukui, G. G. Hall, H. Hartmann, H. H. Jaffé, J. Jortner, W. Kutzelnigg, K. Ruedenberg, E. Scrocco

Springer-Verlag
Berlin
Heidelberg
New York
Tokyo

Lecture Notes in Chemistry

Editors: G. Berthier, M. J. S. Dewar,
H. Fischer, K. Fukui, G. G. Hall,
H. Hartmann, H. H. Jaffé, J. Jortner,
W. Kutzelnigg, K. Ruedenberg, E. Scrocco

Springer-Verlag
Berlin
Heidelberg
New York
Tokyo

Lecture Notes in Chemistry